科技英语翻译

主　编　潘冀春　刘　洁

北京理工大学出版社
BEIJING INSTITUTE OF TECHNOLOGY PRESS

图书在版编目（CIP）数据

科技英语翻译 / 潘冀春，刘洁主编. —北京：北京理工大学出版社，2020.12
ISBN 978-7-5682-9236-8

Ⅰ.①科…　Ⅱ.①潘…②刘…　Ⅲ.①科学技术-英语-翻译-高等学校-教材
Ⅳ.①N43

中国版本图书馆 CIP 数据核字（2020）第 222935 号

出版发行 / 北京理工大学出版社有限责任公司
社　　址 / 北京市海淀区中关村南大街 5 号
邮　　编 / 100081
电　　话 / （010）68914775（总编室）
（010）82562903（教材售后服务热线）
（010）68948351（其他图书服务热线）
网　　址 / http：//www. bitpress. com. cn
经　　销 / 全国各地新华书店
印　　刷 / 三河市天利华印刷装订有限公司
开　　本 / 787 毫米×1092 毫米　1/16
印　　张 / 13.5
字　　数 / 318 千字
版　　次 / 2020 年 12 月第 1 版　2020 年 12 月第 1 次印刷
定　　价 / 36.00 元

责任编辑 / 武丽娟
文案编辑 / 武丽娟
责任校对 / 刘亚男
责任印制 / 施胜娟

前　言

随着我国对外科技交流的日益广泛和深入，科技英语作为交流的工具，愈来愈受到人们的关注。科技英语翻译为拥有不同语言与文化的人们开展科技工作交流架起了一座沟通的桥梁，对科技的进步、发展与传播发挥着巨大的作用。因此，许多高等院校开设的大学英语拓展系列课程当中，“科技英语翻译”均占有十分重要的位置。通过多年的教学实践和研究，我们觉得非常有必要编写一本适合于非英语专业学生，特别是理工科学生使用，且适合一学期教学时长使用的科技英语翻译教材。

本教材共分为科技英语翻译理论概述、科技英语词语翻译、科技英语句子翻译和科技英语篇章翻译四大部分：

第一部分（第1章、第2章）介绍了科技英语的文体特点，阐述了科技英语翻译的基本理论知识，并提出科技英语的翻译标准和对译者的要求。

第二部分（第3章、第4章）介绍了科技英语词语翻译中常用的方法，并重点介绍了科技英语中一些典型词语的翻译技巧。

第三部分（第5章、第6章）首先介绍了句子翻译中具有普遍性的翻译方法，其后通过大量译例，详细阐述了科技英语中一些常见句式和重点句式的翻译方法和技巧。

第四部分（第7章、第8章）从篇章的层面上，首先探讨了中文科技语篇的文体特点，然后揭示了英文科技语篇翻译的基本步骤和翻译要点，最后通过大量实例介绍了不同类型英文科技语篇的翻译策略。

本教材集专业性、知识性和实用性于一体，相比其他同类型的教材，具有以下特点：

1. 教材内容设计简约而有精髓，硬核内容十足；注重理论与实践的结合，突出实践，授人以渔。

2. 教材内容重点突出且层次分明，学习内容的安排循序渐进，由点及面，层层深入。

3. 教材中所选择的译例和篇章，涉及物理、数学、化学、计算机、生物、生命科学、航空、航天、民航等诸多学科领域，适合不同专业学生的学习。

4. 本教材特别设计了“不同类型英文科技语篇翻译实例”这一章节，分门别类地介绍了不同类型英文科技语篇的文体特点和翻译策略。这些语篇类型，除了其他教材中惯常涉及的科普类型、说明文类型之外，还特别关注了诸如学术论文、科技新闻报道、科幻小说以及科教影片字幕等其他教材很少涉及的语篇类型的翻译。该章节学习资源非常丰富，极大地拓展了学生学习的视野。

5. 教材每个章节都设计了大量富有针对性的练习，帮助学生巩固章节知识的学习。练习中还特别增加了“拓展阅读和翻译练习”这一环节，旨在对学生进行篇章翻译的训练，使得学生能够在足够大的语境中，灵活运用所学技巧进行翻译实践，以有效弥补目前很多相关教材重句子、轻篇章，重理论、轻实践的不足。

6. 本教材的译例为编者精挑细选，具有很强的针对性。其中的难点译例，附有详细的翻译分析或常见翻译错误的讲解，以便更好地帮助学生掌握和运用翻译方法和技巧。

本教材适用于非英语专业已过四、六级的本科生，或者具有同等程度的科技英语翻译爱好者自学使用。编写的目的旨在帮助学生学习和了解科技英语的特点，掌握科技英语翻译的基本理论、方法和技巧，在此基础之上，能够利用工具书翻译英语文献，以及一般性的各类科技文章。通过该教材的学习，不仅能切实提高学生英语实际运用能力，而且对学生在未来工作领域中从事科技英语翻译，会起到积极的帮助作用。

编写此书，是对我们多年从事的科技英语翻译教学研究工作的一次总结，鉴于水平和经验有限，书中如有疏漏和不当之处，诚请同行学者和广大读者予以指正。

编　者

目　录

第1章 概述

人们在运用语言的过程中，会根据不同的交际内容、对象、范围、语境和交际目的而采用不同的语言体式，这称为语体（variety）。科技语体属书面文体之一，是准确地记录、描述自然现象，阐发、论证这些现象发生、发展的内在规律，揭示客观事物本质的用语体式。语体是在语境类型作用下的言语功能变体，可以适应各种言语环境的表达要求，具有各种相应的表达功能。各种不同的语体在语言的运用和修辞手法上，都表现出它们各自特有的风格。以下，将介绍科技英语作为科技语体在文体上的主要特点。

第1节 科技英语及其文体特点

科技英语（English for Science and Technology）是英语的科技语体，普遍认为它随着现代科学技术的迅猛发展而诞生于20世纪50年代。刘宓庆先生在谈到科技英语的文体范畴时指出，它可以包括：①科技著述、科技论文和报告、实验报告和方案；②各类科技情报和文字资料；③科技实用手册；④有关科技问题的会谈、会议、交谈的用语；⑤有关科技的影片、录像等有声资料的解说词；⑥科技发明、发现的报道；⑦科学幻想小说等。作为一种独立的语体，科技英语必然具有不同于其他语体的特点，以下将从词汇、语法和修辞方面对科技英语的文体特点加以分析。

1.1 词汇特点

（1）科技英语中大量使用专业术语，其中有一些术语是纯科技术语；还有一些是普通词汇，又称共核词汇，这些词汇在科技语篇里常常具有不同于日常生活中的意义，且在不同学科中含义有时也会有所不同。如 work 在日常生活语中一般指“工作”，而在物理学上则指“功”；frame 一词在机械原理中可作“机架”，而在电信技术中则又作“帧”或“镜头”等。

（2）词形一般较长，且多源于希腊或拉丁语词汇。据美国科技专家 O. E. Nybaken 统计，一万个普通词汇中，约有46% 源于拉丁语，7.2% 源于希腊语，在科技英语中，专业性越强，这种比率越高。此外，源于希腊或拉丁语的前缀、后缀，尤其是源于拉丁语的，有极强的构词能力，以 semi，auto，-able，-ment，-ion 等构成的词随处可见。

（3）多复合词。例如：radiophotography 无线电传真、anti-armoured-fighting-vehicle-missile 反装甲车导弹等。

（4）多缩略词。例如：cpd（compound）化合物、FM（frequency modulation）调频、TELESAT（telecommunications Satellite）通信卫星等。

(5) 多名词化结构。为使行文简洁，表述客观、精确，科技英语中有显著的名词化倾向，即多用能表示动作或状态的抽象名词或起名词作用的动名词形式。例如：the transmission and reception of images of moving objects by radio waves 通过无线电波发射和接收活动物体的图像、illumination intensity determination 照明强度测定、computer programming teaching device manual 计算机程序编制教学装置手册。

1.2 语法特点

(1) 动词时态多用一般现在时以便更好地表现文章内容的无时间性，说明那些科学定义、定理、公式不受时间限制，任何时候都成立，使文章内容显得更客观准确。例如：

There are a number of methods of joining metal articles together, depending on the type of the metal and the strength of the joint which is required. Soldering gives a satisfactory joint for light articles of steel, copper or brass, but the strength of a soldered joint is rather less than a joint which is brazed, riveted or welded. These methods of joining metal are normally adopted for strong permanent joints.

把金属连接在一起的方法很多，因金属的类型和所需接缝的强度不同而异。软钎焊焊接薄钢件、薄铜件和薄黄铜件，其接缝合乎要求，但其强度比硬钎焊铆接或焊接接缝要低得多。这三种连接金属的方法通常用于高强度的永久性焊接。

The elimination of the errors of these types is a major problem in the design of physical experiments. A few possible lines of attack on this problem are indicated below, but the devising of experiments to investigate a particular point, to the exclusion of disturbing factors, is the most exacting of experimental physicist.

消除这几种误差，是设计物理学实验的重要问题。以下叙述了解决这个问题的几种可行办法，不过，要排除各个干扰因素设计出一个实验方案来，去对特定的问题加以研究，却是实验物理学家一项最艰巨的任务。

科技英语中，过去时以及现在完成时多用于科技发展史、科技报告和科技报刊中，用来表述已有的发现或已获得的成果。

(2) 多使用第三人称，尽量避免使用第一或第二人称，以使叙述更显客观，更少主观臆断的色彩。例如：

Gold, a precious metal, is prized for two important characteristics. First of all, gold has a lustrous beauty that is resistant to corrosion. Therefore, it is suitable for jewelry, coins and ornamental purposes. Gold never needs to be polished and will remain beautiful forever. For example, a Macedonian coin remains as untarnished today as the day it was minted twenty-three centuries ago. Another important characteristic of gold is its usefulness to industry and science. For many years, it has been used in hundreds of industrial applications. The most recent use of gold is in astronaut's suits. Astronauts wear gold-plated heat shields for protection outside the spaceship. In conclusion, gold is treasured not only for its beauty, but also for its utility.

黄金这种贵重金属，因两个重要特点而备受人们青睐。首先，黄金具有抗腐蚀的美丽光泽，因此，很适合用来做珠宝、硬币，或者用于其他装饰性用途。黄金从不需要抛光，却能永远保持美丽。例如，一枚马其顿金币即使在今天也依然像它在2300年前刚铸造出来时那样光彩夺目。黄金的另一个重要特点是它对工业及科学大有用处。多年以来，它一直在数百个工业应用中得到使用。最近，黄金甚至应用于航天员的宇航服中。航天员出了宇宙飞船就要穿上镀金的防热罩来保护自己。总之，黄金不仅因为美丽，也因为其实用性而受到人们的珍爱。

（3）大量使用被动语态。

总的来讲，英语使用被动语态的频率远高于汉语，在各种文体中都是如此。而被动语态的使用在科技英语中又比在其他文体中还要多，这是因为科技英语中被描述的主体通常是客观事物或现象，被动语态能较好地服务于科技英语的表述，同时使读者关注所描述的对象。例如：

Consequently, the suggested procedures **must be viewed only as** the best presently available in the opinion of the specialists who worked on this chapter.

因此，根据编写本章的专家们的意见，应将建议的程序看作现有的最好程序。

Attention must **be paid to** the working temperature of the machine.

应当注意机器的工作温度。

The results of the experiment **should be** plotted on a graph.

实验结果应在图上标出。

The surface **must be** cleaned.

表面必须清扫干净。

Flux **must be** applied.

必须涂上助熔剂。

（4）大量使用非谓语动词。

英语中的非谓语动词包括现在分词、过去分词和不定式三种。科技文章行文简练、结构紧凑，为此，往往使用分词短语代替定语从句或状语从句。这样可缩短句子，又比较醒目。例如：

A direct current is a current **flowing** always in the same direction.

直流电是一种总是沿同一方向流动的电流。

We can store electrical energy in two metal plates **separated** by an insulating medium.

电能可储存在由一绝缘介质隔开的两块金属极板内。

A body can move uniformly and in a straight line, **there being** no cause to change that motion.

如果没有改变物体运动的原因，那么物体将作匀速直线运动。

（5）大量使用后置定语。

大量使用后置定语也是科技英语的特点之一，目的是简化句子结构和对被修饰词进行严格限定和说明。常见的后置定语结构有以下几种：

介词短语

The forces **due to** friction are called frictional forces.

由于摩擦而产生的力称之为摩擦力。

A call **for paper** is now being issued.

征集论文的通知现在正陆续发出。

形容词及形容词短语

In this country the only fuel **available** is coal.

该国唯一可用的燃料是煤。

In radiation, thermal energy is transformed into radiant energy, **similar in nature to light**.

热能在辐射时，转换成性质与光相似的辐射能。

副词

The air outside pressed the side **in**.

外面的空气将桶壁压得凹进去了。

The force **upward** equals the force **downward** so that the balloon stays at the level.

向上的力与向下的力相等，所以气球就保持在这一高度。

非谓语动词

The results **obtained** must be checked.

得到的结果必须加以校核。

The heat **produced** is equal to the electrical energy wasted.

产生的热量等于消耗了的电能。

定语从句

During construction, problems often arise **which require design changes**.

在施工过程中，常会出现需要改变设计的问题。

The molecules exert forces upon each other, **which depend upon the distance between them**.

分子相互间都存在着力的作用，该力的大小取决于分子之间的距离。

Very wonderful changes in matter take place before our eyes every day **to which we pay little attention**.

我们几乎没有注意到，很奇异的物质变化每天都在眼前发生着。

（6）大量使用祈使句。

祈使句是表示请求、命令、建议、劝阻等的句子，具有表达简单明了、干脆利落的特点，因此，广泛运用于科技英语中的特定语篇中，例如使用说明、实验步骤等的描述中。

Allow the water to cool for ten minutes and then take the temperature.

让水冷却10分钟，然后再测温度。

Grip handles as far as possible from blades.

紧握手柄，尽量离刀口远些。

Handle the machines with great cares.

这些机器应小心轻放。

(7) 长句较多。

一般而言，句长在15个词以下的视为短句，而根据统计资料表明，现代科技英语作品的平均句长在20~30个词之间。科技英语的作用是陈述事理、描述过程，它所给出的定义、定理、定律或描述的概念或工艺过程，都必须严谨、精确，因而科技英语逻辑性强、结构严密、表达明确。这些特点规定科技英语必然带有许多修饰、限定和附加成分。这些成分一多，句子必然就长了、复杂了。科技英语中的长句可能是含有较多修饰成分的简单句，也可能是并列句，或者是主从复合句。例如：

The general layout of the illumination system and lenses of the electron microscope corresponds to the layout of the light microscope.

电子显微镜的聚光系统和透镜的设计与光学显微镜的设计是一致的。

上面例子中这个长句是一个较长的简单句，而以下这个长句则是较复杂的并列句，其中一个分句包含条件状语，而另一个并列分句则包含了一个原因状语：

A computer cannot do anything unless a person tells it what to do and gives it the appropriate information; but because electric pulses can move at the speed of light, a computer can carry out vast numbers of arithmetic-logic operations almost instantaneously.

如果人们不给予指示，也不提供适当的信息，计算机便什么也做不了。但是因为电子脉冲能够以光速运动，所以计算机能够瞬间就执行大量算数-逻辑运算。

以下这个长句则是包含后置定语从句的主从复合句：

The area rule concept, the supercritical airfoil, and the engine installation configurations that provide favorable interference which are presently being applied to military aircraft and fuel-efficient transport aircraft were evolved through wind tunnel experiments.

面积律概念、超临界翼型、可提供有利干扰的动力装置布局等，都是通过风洞实验发展起来的。后者现在正应用于军用飞机及低油耗运输机上。

一般长句的成分之间存在简单的线性序列关系，而多重复句的深层结构，即内在的语法关系比较复杂，更能有效地表达严谨细致的复杂思维活动。因此，科技英语中的多重复句被认为是最能表现其语言风格的重要语法现象之一。例如：

In this way the distinction between heavy current electrical engineering and light current electrical engineering can be said to have disappeared, but we still have the conceptual difference in that in power engineering the primary concern is to transport energy between distant points in space; while with communications systems the primary objective is to convey, extract and process information in which process considerable amounts of power may be consumed.

分析：

上面的多重复句从总体上说是一个并列复合句，两个并列分句由but连接，表示转折。并列句的第一分句是个简单句，而第二分句是个主从复合句，其中 in that 引导一个方面/原因状语从句，而在该从句中又有两个分句之间存在并列关系，由 while 连接表示对比；while 之后的分句中又包含一个由 in which 引导的限制性定语从句，但它在语意上却是一个表示转折关系的分句，来修饰非谓语动词短语所表示的过程。

译文：

在这方面，强电工程和弱电工程之间的区别可以说已经消失；但是我们仍旧认为它们在概念上有所不同，因为电力工程的主要任务是在空间相距较远的各地之间输送能量，而通信系统的主要目的则是传递、提取和处理信息，尽管在这个过程中或许消耗相当大的电力。

1.3 修辞特点

科技英语以传递信息为主要目的，其语言风格准确、客观、逻辑性强，这必然使其具有不同于文学语体的修辞特点。例如以下两段文本：一个是文学文本，一个是科技文本：

She was of a helpless, fleshy build, with a frank, open countenance and an innocent, diffident manner. Her eyes were large and patient, and in them dwelt such a shadow of distress as only those who have looked sympathetically into the countenances of the distraught and helpless poor know anything about. (from *Jennie Gerhardt*)

那妇人生着一副绵软多肉的体格，一张坦率开诚的面容，一种天真羞怯的神气。一双大落落的柔顺眼睛，里边隐藏着无穷的心事，只有那些对于凄惶无告的穷苦人面目作过同情观察的人才看得出来。

——摘自《珍妮姑娘》((*Jennie Gerhardt*), 傅东华译)

分析：

以上片段摘自小说《珍妮姑娘》，仅仅 51 个词的片断，就运用了 10 个形容词，占五分之一。

The programmer can concentrate on the general program logic, without becoming bogged down in the syntactic details of the individual instructions. In fact, this entire process might be repeated several times, with more programming detail added at each stage. Once the overall program strategy has been clearly established, then the syntactic details of the language can be considered. Such an approach is often referred to as "top-down" programming.

程序员可将注意力集中在总的程序逻辑上而不会陷入各条指令的句法细节中。事实上整个过程可能重复若干次而每个阶段都要加入更多的编程细节。一旦清楚地制定了全面的编程策略便可以考虑该语言的句法细节。这种方法经常称作"自顶向下"的程序设计法。

分析：

从上面的这段科技文章，不难看出其文体与前面的文艺小说迥然不同。总体而言，科技文体严谨周密，概念准确，逻辑性强，行文简练，重点突出，句式严整，少有变化。文学语言常运用诸如比喻、拟人、夸张等各种修辞手段，来增强语言的表现力与感染力，从而给读者新奇的阅读体验；而科技语言则极少运用这些修辞手段，因为运用这些修辞手段与其语言精确、平实、客观公正的风格相悖。因此，科技英语的修辞特点就是修辞单调，甚至是没有常规意义的修辞。

科技英语作为英语文体中的一种，也具有普通英语的一些普遍特点。因此，普通英语的一些翻译技巧和策略也适用于科技英语的翻译。另一方面，掌握了科技英语的上述基本特点，可以帮助译者更有效而且准确地将科技文献译成汉语。

第2节 翻译与科技英语翻译

翻译实践历史久远，为不同语言文化的人们架起了沟通与交流的桥梁。但到底什么是翻译，不同的视角会有不同的界定。有人认为翻译是一门科学，因为它有着自己的内在科学规律；也有人视翻译为一门艺术，因为翻译好比作画，先抓住客观世界中人或物的形态和神态，再用画笔将其惟妙惟肖地表现在画上；还有人将翻译认作是一门技能，因为就其具体操作过程而言总是离不开方法和技巧。美国翻译理论家奈达将其定义为“用自然贴切的译入语将源语信息再现出来，并且在内容与风格上要达到对等。”而英国理论家卡特福德则这样界定，“翻译是将一种语言（源语）的文本材料用另一种语言（译入语）对等的文本材料予以替换。”而文学翻译家则倾向于将其视为一种再创作。

不论如何界定翻译，这里有几个概念需要掌握：译出语或者源语，指的是待译文本原来的书写语言；译入语或者目的语，指待译文本所要转换成的语言。多数翻译理论家都强调了翻译必须达到对等的效果。对等指的是，译文与原文要在内容与风格上达到同等或类似的效果，译文读者读了后就会产生类似于源语读者读了原文后的体验。只有达到这样的效果，才算是成功的翻译。

既然翻译可从不同的角度来定义，那么同样，翻译的种类也可从不同的视角来分类。一般说来，翻译可从5种不同的角度来分类：

（1）从译出语和译入语的角度来分类，翻译可分为本族语译为外语、外语译为本族语，例如：中译英、英译中。

（2）从涉及的语言符号来分类，翻译分为语内翻译（intralingual translation），例如，文言文译为现代汉语；语际翻译（interlingual translation），即通常意义上的翻译；符际翻译（inter-semiotic translation）是一种广义的翻译，例如，你看到不同的交通信号灯或者不同的标示牌，就会做出相应的举动，这就是一种符际翻译。

（3）从翻译的手段来分类，翻译可分为口译（interpretation）、笔译（interlingual translation）和机器翻译（machine translation，指借助翻译软件来完成的翻译）。

（4）从翻译的题材或者内容来分类，翻译可分为专业文献翻译（translation of English for Science and Technology），即所谓的科技英语翻译；文学翻译（literary translation）；一般性翻译（practical writing translation），除以上两类文本之外任何类型文本的翻译，均属于一般性翻译，又称通用翻译。

（5）从翻译的处理方式来分类，翻译可分为全译（full translation），指将待译文本的所有内容从头到尾翻译出来，没有内容上的删减；摘译（partial translation）则是根据客户的需求，选择其需要的相关内容进行翻译；编译（translation plus editing）是将待译文本的内容用译入语大致表达出来，可以进行必要的编辑工作，也就是对原文的内容有所增删。

由此可见，科技英语翻译是特定类别的翻译，是基于待译文本题材或者内容所做的划分。从译出语与译入语的角度来看，属于英译汉，而从翻译处理的手段来看是属于笔译。好的翻译必须在内容与风格上达到与原文同等的效果，那么，科技英语具有的客观、准确以及严密的逻辑性的特点在译文中必须如实传达，切勿将一个科技文本译成了其他文体的文本。

第3节　翻译的本质与过程

科技英语翻译作为翻译的一个类别，其本质及翻译过程与一般的翻译并无二致。总体而言，翻译本质上是理解与表达的过程，也是解码与编码的过程，整个过程分为理解、表达、校核三个阶段。

（1）理解的过程即是对源语解码的过程，可以是指广义的理解，这一过程具有社会性，因此不能忽视社会效应对这一过程的结果所带来的全面影响。广义理解包括对原文作者、原文产生的时代背景、作品的内容以及原文读者对该作品的反映的理解。狭义的理解仅指对待译文本的理解，主要包括语法分析、语义分析、语体分析和语篇分析。理解是翻译成功与否的先决条件和重要步骤，务必正确可靠，杜绝谬误。没有正确的理解，翻译就失去了意义。例如：

Prefabrication is by no means a new idea, but in the past, unimaginative uniformity in design, together with limited materials, led to a natural distaste for this form of construction. Recent years, however, have produced a revolution in this field, as a result of new techniques, wider ranges of materials and a far more variety in the shapes and sizes of prefabricated units. In fact, nowadays, there remains very little in the world of building and civil engineering that cannot be constructed effectively and attractively by means of prefabricated units.

Generally speaking, prefabricated construction in relation to building is divided into two classes. When the prefabricated units are entirely or almost entirely produced in a factory and transported to the site, the method of construction is known as factory industrialized building or system building. On the other hand, if the component parts are mainly produced by the contractor in a yard adjacent to the site, it is known as on-site industrialized building.

Many large building contractors have developed their own organizations of system building for houses and flats, and have often been successful in securing contracts in direct competition with traditional building.

The main advantage of factory industrialized building is that the prefabricated units can be produced to a very high standard of precision and a consistently highquality. Consequently, the tolerance margins, which are so necessary in traditional building methods, can be greatly reduced. Of course, careful planning and diligent supervision by the designer, together with the utmost standardization, are essential to attain these objectives. The working conditions in a factory are far more amenable than those on a site, and this permits continuous production. Since unskilled labor is mainly employed in the production of the units, construction should prove to be less costly.

以上这段文字主要涉及预制建筑法的发展、分类以及主要优点，其中一个关键术语 prefabricated units（预制构件）在文中多次出现，有时是以相同形式，而有时却是以不同形式出现，例如在第二段中是以近义词 component parts 的形式出现，而在最后一段中又变成了上义词 the units。如果译者不能结合上下文正确理解该术语的不同形式，而是受字面意思影响，将其翻译成“组成部件”“单元”或者“单位”，这无疑会因为理解上的谬误而造成误译，进而误导读者。如果没有正确的理解，就算译文表达得多么流畅，也是枉然。因此，在阅读待译文本时，务必要保证理解正确。

（2）表达是理解后能否保证译文成功的又一关键步骤，是理解的深化和体现。倘若译者能正确理解原文，但表达不出所理解的内容，那么，理解得再正确也是枉然。因此，译者必须在选词用字、组词成句、组句成篇上下功夫，在翻译技巧的运用上下功夫。表达时还应注意避免翻译腔、过分表达和欠表达三种情况。

翻译腔，就是指译文不符合汉语语法和表达习惯，佶屈聱牙，晦涩难懂。换言之，就是表达不地道。例如：

Prefabrication is by no means a new idea, but in the past, unimaginative uniformity in design, together with limited materials, led to a natural distaste for this form of construction.

原译：

预制绝不是一个新想法，但在过去，缺乏想象力的统一设计/缺乏普遍的想象力的设计，导致对这种材料的自然厌恶。

分析：

在这句话的翻译中，“缺乏想象力的统一设计/缺乏普遍的想象力的设计”是对原文中名词化结构 unimaginative uniformity in design 的错误理解导致的误译，而“自然厌恶”又是对原文字对字硬译的结果，导致浓重的翻译腔。而“想法”一词则过于口语化，与科技文本的风格不符。

改译：

预制绝非是现在才有的理念。但在过去，由于设计缺乏想象力，千篇一律，再加上材料有限，自然使这种建筑方法不受人们青睐。

过分表达，就是指译文画蛇添足，增加了原文没有的内容。例如：

Recent years, however, have produced a revolution in this field, as a result of new techniques, wider ranges of materials and a far more variety in the shapes and sizes of prefabricated units.

原译：

然而，近年来由于新技术的出现，在世界建筑和土木工程领域中，范围更广、种类更多的材料仍然很少，无法通过预制构件有效地施工，因此，这一领域产生了革命。

分析：

在译文当中，增添了原文没有的内容，诸如“范围更广、种类更多的材料仍然很少，无法通过预制构件有效地施工”，显然是译者没有真正看懂原文，于是离开原文随意发挥，擅自增加自以为合理的解释的结果。

改译：

但近年来，由于新技术的出现，材料选择范围更加广泛，以及预制构件形状和规格的多样化，使这一领域发生了一场变革。

欠表达，则是指译文省略或删节原文的内容。例如：

The main advantage of factory industrialized building is that the prefabricated units can be produced to a very high standard of precision and a consistently high quality.

原译：

工厂工业建筑的主要优势是可以生产精准度高、质量高的预制构件。

分析：

原译未能表达出“始终如一的高质量”这个意思。

改译：

工厂预制法的主要优点是，生产出的预制构件具有很高的精度标准和始终如一的高质量。

(3) 校核是对理解和表达质量的全面检查，是纠正错误、改进译文的极好时机，切不可认为是多余之举。优秀的译者总是十分重视校核的作用，总是利用这一良机来克服自己可能犯下的错误，初学翻译的人就更应该如此了。在校核时要注意检查译文的格式、文字输入和拼写、标点错误；数字有无错漏；人名、地名是否统一；句子、段落有无漏译等。校核一般要进行二至三遍。第一、二遍，先对照原文，检查内容上有没有漏译或者误译的地方；第三遍，可以脱离原文，仅看译文来润饰文字。只有这样，才能尽可能保证译文的质量。

第 4 节　练习

一、思考题

1. 科技英语作为科技语体在词汇、语法、修辞上分别具有什么特点？
2. 科技英语翻译与翻译有何关联？
3. 翻译过程具有哪三个阶段？其相互关系是什么？

二、翻译实践

阅读以下段落并体会科技英语的各个特点，之后翻译以下段落，再将你的译文与参考译文比较，尤其注意划线部分的处理。

Fiber-optic lines will form most of the backbone of the information highway, just as they do for the phone system today. Fiber-optic cable is made of long, thin strands of glass rather than wire, and it transmits information in the form of digitized pulses of laser light rather than the radio waves used by coaxial cable. Because light pulses have shorter wavelengths than radio waves, engineers can cram much more data into fiber-optic lines than into other kinds of cables and wires.

A single fiber, for example, can handle a mind-boggling 5,000 video signals or more than 500,000 voice conversations simultaneously. This huge capacity allows it to transmit all signals digitally. So noise or static easily can be filtered out. Finally, because glass is an inherently more efficient medium for transmitting information than other materials, a fiber-optic line can transmit a signal thousands of miles without much "signal loss". Fiber-optic cable, simply put, is the method of choice for transmitting massive quantities of information over long distances.

Another key is "digital compression" —a variety of methods for reducing the amount of digital code (streams of ones and zeros) needed to represent a piece of information—whether it is a document, a still picture, a movie or a sound. Digital compression is most critical for transmitting video, because digitized video consumes enormous amounts of space. Just four seconds of a digitized film, for example, would completely fill a 100-megabyte hard drive. A feature film of typical length, if uncompressed, would occupy more than 350 ordinary compact discs.

Compression techniques achieve their gains by recording only the changes from one frame to the next. The background image in a movie scene, for example, typically does not change much from one frame to another. In a digital compression scheme, the background would be recorded only once; after that, only the actors' movements would be captured.

One result is more choices—hundreds of channels coming through your cable TV line instead of dozens. Digital compression also makes it easier to piggyback data onto media that were not designed with data in mind: in particular, phone lines.

参考译文：

光纤电缆构成了今日电话系统的主干线，同样，它也将构成未来信息公路的主干线。光纤电缆的制造材料是细长的玻璃丝而不是金属导线，它的信息传输方式是激光数字脉冲而不是同轴电缆使用的无线电波。由于光脉冲的波长比无线电波的波长短，因而工程师可以输入光纤电缆的信息量之大令其他类型的电缆和导线望尘莫及。

例如，一根光纤电缆可同时处理5 000个电视信号或50多万次电话通话，简直令人惊奇。这种巨大的能力使它可以实现全部信号传输数字化，从而轻而易举地滤掉噪声或静电干扰。最后要指出的是，玻璃作为媒介，它本身的传输效率便高于其他材料，因而一条光纤电缆可将一个信号传输数千英里而几乎没有“信号流失”。简而言之，光纤电缆是远距离传输大量信息的最佳方式。

另一关键技术是“数字压缩”技术，即减少数码（一连串的1和0）数量的各种方法，这种数码用来表示某种信息，如一份文件、一张静止图片、一部电影或一种声音。数字压缩技术对影像传送来说至关重要，因为数字化影像会占据大量的存储空间。例如，仅仅四秒钟的数字化影片，便要占据一百兆的硬盘存储空间。一部普通长度的故事片，如果未经压缩，要占350多张普通光盘。

数字压缩技术只记录一个画面与下一个画面之间的变化，从而实现其功能。例如，电影场景的背景图像在一个画面转为另一个画面时通常是不变的，在数字压缩系统中，背景图像只被记录一次，此后便只记录演员的动作。

数字压缩技术使可供选择的有线电视频道从几十个增加到几百个，数字压缩技术还能使数据更容易通过本不是用于传送数据的媒体线传送——特别是电话线。

三、拓展阅读和翻译练习

阅读以下语篇，并翻译划线句子及黑体段落。

In Italy about three hundred years ago there lived a young man whose name was Galileo. Like Archimedes he was always thinking and always asking the reasons for things. He invented the thermometer and simple forms of the telescope and the microscope. He made many important discoveries in science.

One evening when he was only 18 years old, he was in the cathedral at Pisa at about the time the lamps were lighted. The lamps—which burned only oil in those days—were hung by long rods from the ceiling. 1. When the lamplighter knocked against them, or the wind blew through the cathedral, they would swing back and forth like pendulums. Galileo noticed this. Then he began to study them more closely.

He saw that those which were hung on rods of the same length swung back and forth, or vibrated, in the same length of time. Those that were on the shorter rods vibrated much faster than those on the longer rods. As Galileo watched them swinging to and fro he became much interested. 2. Millions of people had seen lamps moving in this same way but no one had ever thought of discovering any useful fact connected with the phenomenon.

When Galileo went to his room he began to experiment. He took a number of cords of different lengths and hung them from the ceiling. 3. To the free end of each cord he fastened a weight, then he set all to swinging back and forth, like the lamps in the cathedral. Each cord was a pendulum, just as each rod had been.

He found after long study that when a cord was 39.1inches long, it vibrated just 60 times a minute. A cord one-fourth as long vibrated just twice as fast, or once every half second. To vibrate three times as fast, or once in every third part of a second, the cord had to be only one-ninth of 39.1inches in length. By experimenting in various ways Galileo at last discovered how to attach pendulum to time pieces as we have them now.

4. Thus, to the swinging lamps in the cathedral, and to Galileo's habit of thinking and inquiring, the world owes one of the commonest and most useful inventions—the pendulum clock.

You can make a pendulum for yourself with a cord and a weight of any kind. You can experiment with it if you wish; and perhaps you can find out how long a pendulum must be to vibrate once in two seconds.

Source：《大学英语四六级晨读经典：夏日展望篇》

第2章　科技英语翻译的标准

第1节　翻译的标准

翻译的标准，指在翻译实践中译者所遵循的翻译准则。这些准则不仅对译者的翻译实践活动起指导作用，而且也是用来衡量译文优劣的尺度和标准。切实可行的翻译标准对发挥翻译功能、提高翻译质量具有重要的意义。

古今中外，翻译实践者和学者都提出过不少翻译标准。其中，影响最大的当属严复的“信、达、雅”三字标准。清末的启蒙思想家、教育家、翻译家严复于1897年在《天演论》的序言中提出了“信、达、雅”之说。所谓“信”，就是译文的思想内容忠实于原作；“达”就是译文不拘泥于原文形式，力求表达通顺明白；“雅”是指追求译文的古雅，即要求风格优美。但严复对于“雅”的追求是“与其伤雅，毋宁失真”，这个观点在今天看来失之片面。后来，鲁迅在《且介亭杂文二集》提出了“信顺”之说的翻译标准。他说：“凡是翻译，必须兼顾着两面，一当然力求其易解，二则保存着原作的风姿”。傅雷在《高老头》重译本序中提出了“神似”之说；钱钟书提出了“化境”之说；刘重德提出了“信、达、切”之说等。

在国外，早在1540年，法国学者埃缔延·多勒就在其《如何出色地翻译》（How to Achieve a Good Translation from One Language into Another）一文中提出了五条翻译原则：

（1）The translator must have a perfect understanding of his author's message and material.（译者必须很好地理解原作的内容和原作者的意图。）

（2）He should have complete mastery of both source and target language.（译者必须具备很好的原文语言知识和译文语言知识。）

（3）He should not translate word for word.（译者应避免逐词翻译。）

（4）He should be aware of Latinisms and use idiomatic language.（译者应注意拉丁语表达并使用地道的语言进行表达。）

（5）He should strive after a smooth, elegant, unpretentious and even style.（大体上保持原文的风格和韵味，避免矫揉造作。）

1790年，英国的泰特勒在其《论翻译的原则》（Essay on the Principles of Translation）一书中也提出了较为系统的翻译“三原则”：

（1）A translation should give a complete transcript of the ideas of the original work.（译文必须能完全传达出原作的意思。）

（2）The style and manner of writing should be of the same character as that of the original.（译文的风格和手法应和原文的性质相符。）

（3）A translation should have all the ease of the original composition.（译文应与原文一样流畅。）

此外，美国著名翻译理论家奈达提出了“动态对等”的翻译原则，进而又从社会语言学和语言交际功能的观点提出了“功能对等”的翻译原则。纽马克也提出了“交际翻译”和“语义翻译”的两大翻译原则。

应该说，每个翻译标准都是在一定的时代或环境中提出来的，没有一个“放之四海皆准”的翻译标准。目前，当代翻译界普遍认同的翻译标准是：忠实、通顺。就是译文要力求确切地表达原文的内容和风格，同时在形式上又符合汉语的表达，做到通顺、流畅、易懂。“忠实”是通顺的基础，译文如果不“忠实”，再“通顺”也失去了意义。反过来，译文如果不通顺又必然影响到译文的“忠实”。所以一篇好的译文一定要做到“忠实”与“通顺”相结合，即用规范的译文语言形式，准确完整地表达原文的意思，这是对译文质量的基本要求。要达到这个基本要求就必须抓住原文的真正含义，但不要过多地受原文结构形式的束缚，而是按照汉语习惯用法来表达译文。当然，在不影响“忠实”与“通顺”的前提下，若能兼顾原文和译文形式上的统一，则更为理想。

第2节　科技英语翻译的标准

科技英语有别于日常英语和文学英语，其语言规范、语气正式、陈述客观、逻辑性和专业性强。科技英语翻译的目的是将科技英语文字中所表述的理论、概念、观点、方法、结论等内容准确无误、表达清楚、合乎标准地译成汉语以供教学、科研、生产等方面的应用。因此，根据科技英语的自身特点和科技英语翻译的要求，科技英语翻译应遵循忠实、通顺和规范这三个标准和原则。

（1）忠实（Faithfulness）。

所谓忠实，就是指译文忠实于原文。要求译者准确地、不折不扣地传达原文的全部信息内容，不能有任何的歪曲、遗漏或其他的删改。要做到这一点，译者必须充分地理解原文所表述的内容，其中包括对原文词汇、语法、逻辑关系和科学内容的理解。科技翻译的任何错误，甚至是不准确都会给科学研究、学术交流、生产发展等带来不良影响或巨大损失，甚至是灾难。例如：

For the balance of the section, let's speak in straightforward and elementary terms to describe the function of the electronic computer.

原译：

为使这一节平衡起见，我们将用简单明了的术语来描述这种电子计算机的功能。

分析：

英语中，一词多义的情况相当多，词义选择依赖于上下文，根据具体语言环境而定。原文中的balance一词在句中并不是作为“平衡”之意，而是“剩余部分，其余”的意思。因此，翻译时不但要注意词语直接的、表面的、一般的意义，还要探索其引申的、内涵的、特定的意义。如果只知其一，不知其二，就会造成错误理解。

改译：

在本节的其余部分，我们将用简单明了的术语来描述这种电子计算机的功能。

Velocity changes if either the speed or the direction changes.

原译：

假如力的大小或方向改变了，速度跟着要变化。

分析：

此句将 speed 译成“力的大小”，是一个明显的误译，从而造成整个句意的错误。speed 在此处应该译为速率。在物理学中，速度和速率是两个不同的概念。速度 (velocity) 是矢量，有大小和方向；而速率 (speed) 是标量，有大小而没有方向。另外，原文中 either…or, 表示两者居其一的意思，在这里具有强调作用，因此在译文中也应着重体现出来。

改译：

如果（物体运动的）速率和方向有一个发生变化，则（物体的）运动速度也随之发生变化。

Sulphur dioxide is given off when the sulphide ore of metals are treated in order that the metal may be obtained.

原译：

处理金属的亚硫酸盐矿物时释放出二氧化硫，以便获得金属。

分析：

译文忠实于原文，不仅是词语意思层面上的忠实，也包括语法结构层面上的忠实。此句翻译的主要问题是译者误判了“in order that the metal may be obtained”这一结果状语从句和前面的修饰关系，造成了整个句子理解的混乱。实际上，本句的主句是“Sulphur dioxide is given off”, 而“when the sulphide ore of metals are treated in order that the metal may be obtained”是状语从句部分，其中 in order that 引出的是从句中的结果状语，修饰前面的“the sulphide ore of metals are treated”. 另外，sulphide 应为硫化物，在这里被误译成了亚硫酸盐。

改译：

对金属硫化物矿石进行处理以获得金属时，会释放出二氧化硫。

（2）通顺（Expressiveness）。

所谓通顺，是指译文的语言通顺易懂，符合汉语的语法结构及表达习惯，容易为读者所理解和接受。也就是说，译文语言须明白晓畅、文理通顺、结构合理、逻辑关系清楚，没有死译、硬译、语言晦涩难懂的现象。在翻译过程中，人们经常会碰到虽然能够知悉、理解原文的意思，但译成汉语后却语句不通顺、意思不明确的情况。这是因为在翻译时，有时我们不能很好地从英语表达模式里转换到汉语表达模式中来，译语会受到原语某种程度上的干扰。例如：

Because of the durability of the metal, we know a lot about bronze artifacts.

原译：

这种金属的经久性使我们知道了许多关于青铜器的情况。

分析：

上句的译文基本上采取了字对字翻译，虽然并不存在误译的情况，但整个句子晦涩难懂，概念不清。其实，通过上下文语境分析，我们知道 the metal 指的就是 bronze artifacts, 但是上面译文却没有将“这种金属”和“青铜器”两者进行关联，从而造成不达意的现象。

改译：

我们对于青铜器了解甚多，是因为（青铜器）这种金属制品能保存多年而不损坏。

In particular cases no method of noise reduction treatment may be effective in reducing the noise level at the operator's point to an acceptable level and ear protector may be the only answer.

原译：

某些特殊情况下，如果没有有效的减噪方法把操作者工作地的噪声降到一个可以接受的程度，那么护耳器可能是唯一的答案。

分析：

此句中将 answer 译成“答案”一词和前面的意思不连贯。通过分析，我们发现 answer 和前面提到的 method 表达的是同一个概念，只是换了一种表达而已。因此，汉译时我们也要找到在汉语中能够替代 method(方法) 一词的同义表达。

改译：

某些特殊情况下，如果没有有效的减噪方法，可以将操作者工作场地的噪声降到一个可以接受的程度，那么采用护耳器可能是唯一有效的减噪措施。

The engine has given a consistently good performance.

原译：

这台发动机一直给出好的性能。

分析：

译文中将 has given 直译成“给出”，不符合汉语的表达习惯。

改译：

这台发动机一直工作良好。

（3）规范（Appropriateness）。

基于科技语言的特点和用途，进行科技英语翻译，在做到忠实于原文、表达通顺流畅的基础上，我们还应做到规范，即译文必须要使用规范的科技语言和科技术语来进行表达，以体现专业性的要求。如果译文采取普通的表达方式，即便译句意思无误，也是不可取的。例如：

By the age of 19 Gauss had discovered for himself and proved a remarkable theorem in number theory known as the law of quadratic reciprocity.

原译：

高斯十九岁时已经独立地发现并证明了数理定理——二次互反律。

改译：

高斯十九岁时已经独立地发现并证明了数论中的一个卓越定理，名为二次互反律。

科技英语翻译的“规范性”也体现在科技英语译文“简洁明晰”这一特点上。例如：

The dead-soft condition of aluminum foil is the one best suited for the most packaging applications.

原译：

铝箔的极软状态是最适合大多数包装用途的状态。

改译：

铝箔极软，最适于作大多数物品的包装材料。

Very pure copper is need for the wires used in electrical engineering, because quite small amounts of some impurities make copper a much poor conductor of electricity.

原译：

电工中应用的导线需要很纯的铜来制造，因为相当少量的某种杂质都会使铜成为差得多的导电体。

改译：

电气导线需要用很纯的铜，因为极少量的杂质都会使其导电性大大下降。

在科技英语翻译中，忠实、通顺和规范三大翻译标准和指导原则不是孤立存在的，因此不能片面地去对待。在翻译实践中，只有真正做到三者兼顾，才能高质量地完成翻译工作。

第3节 从事科技英语翻译应具备的素质

要做好科技翻译工作，成为一个合格的科技翻译人员，需具备和培养一些基本的素养。

（1）恪守认真负责、一丝不苟、考据求证、译风严谨的治学态度。从事科技翻译工作，我们首先要热爱科技翻译事业，坚决反对和克服粗枝大叶、不求甚解、马马虎虎的翻译作风。严肃认真不仅是一个翻译态度问题，同时也是一个职业训练问题。在平时翻译实践和训练中，要做到勤思考、勤查考、勤积累、勤比较、勤商讨，不断提高自己的翻译水平。

（2）熟练掌握英汉两种语言。首先，具有扎实的英语基础和较高的英语阅读能力，能准确地理解英语原文是保证翻译质量的一个基本条件，因此译者需要不断提高自己的英语水平，充分了解科技英语的特点，为熟练、准确、迅速地进行阅读和翻译打下良好的基础；另外，从事科技英语翻译还要求我们具有扎实的汉语语言表达能力，因为汉语水平的高低是决定翻译质量的另一大要素。因此，译者需要下功夫不断提高自己中文语言文字素养。此外，在平时的训练中，译者还应经常对英汉两种语言的科技词汇、句型结构、表达方式等进行对比研究，以更好地掌握两种语言转换的规律和机制，从而促进翻译能力和水平的提升。

（3）具备良好的科学素质和丰富的科技知识。从事科技翻译，若不懂得所译的专业内容，就会影响理解。因此，为了更好地完成翻译工作，译者要不断吸收和丰富自己的科学文化和专业技术知识，只有对原文涉及的内容了解得越多、知识越丰富，对原文的理解才越深刻，译文的表达也就越准确、越到位。

（4）熟悉相关的翻译理论、掌握常用的翻译方法和技巧。从事科技翻译工作，掌握并能熟练运用翻译理论和翻译方法，做到理论和实践相结合，同样是对科技英语翻译工作者必不可少的要求之一。涉及这部分的相关内容是本书后面的章节主要探讨的内容，希望大家认真学习和掌握。

（5）要有一定的政治素养。自然科学文献是讲述客观事物的科学道理和客观规律的。但有些英语科技文章或报刊有时会涉及鼓吹自己的政治观点，甚至攻击、诽谤我国的国家制度和政策，伤害我们的民族感情等方面的内容。翻译时遇到此类情况，我们一定要予以高度的重视，对这些内容，应当加以正确处理和说明。

第4节　练习

一、思考题

1. 科技英语翻译所遵循的标准是什么？科技英语翻译的标准与广义上的翻译标准有什么异同之处？

2. 从事科技英语翻译应具备哪些方面的素养？结合自己的学习经历，谈谈应该如何提高科技英语翻译工作者的专业素质。

二、翻译实践

翻译下面的句子，注意在翻译过程中体会科技英语翻译的“三个标准”的指导意义和作用。

1. Oil and gas will continue to be our chief source of fuel.
2. The power plant is the heart of a ship.
3. Burning in the cylinder, oil provides the engine with power.
4. When a person sees, smells, hears or touches something, then he is perceiving.
5. From these discoveries, the scientists concluded that the woman corpse died of pneumonia.
6. Heat from the sun comes to us by radiation.
7. In modern reciprocating steam engine, condensation problems have been practically eliminated.
8. If we close our eyes, we cannot see anything because our eyelids prevent the rays from entering our eyes.
9. All living things must, by reason of physiological limitations, die.
10. The Red Cross has begun a major cloning project relating to the reproduction of transgenic pigs for organ donor.

三、拓展阅读和翻译练习

阅读以下语篇，并翻译划线句子及黑体段落。

1. As motorways become more and more clogged up with traffic, a new generation of flying cars will be needed to ferry people along skyways. That is the verdict of engineers from the US space agency and aeronautical firms, who envision future commuters travelling by “skycar”. These could look much like the concept skycar shown in the picture, designed by Boeing research and development.

However, such vehicles could be some 25 years from appearing on the market. Efforts to build flying vehicles in the past have not been very successful. 2. Such vehicles would not only be expensive and require the skills of a trained pilot to fly, but there are significant engineering challenges involved in developing them.

"When you try to combine them, you get the worst of both worlds: a very heavy, slow, expensive vehicle that's hard to use," said Mark Moore, head of the personal air vehicle (PAV) division of the vehicle systems program at NASA's Langley Research Center in Hampton, US.

3. But Boeing is also considering how to police the airways and prevent total pandemonium if thousands of flying cars enter the skies.

"The neat, gee-whiz part is thinking about what would the vehicle itself look like," said Dick Paul, a vice president with Phantom Works, Boeing's research and development arm. "But we're trying to think through all the ramifications of what it would take to deploy a fleet of these."

4. Past proposals to solve this problem have included artificial intelligence systems to prevent collisions between air traffic. NASA is working on flying vehicles with the initial goal of transforming small plane travel. 5. Small planes are generally costly, loud, require months of training and lots of money to operate, making flying to work impractical for most people. But within five years, NASA researchers hope to develop technology for a small plane that can fly out of regional airports, costs less than $ 100,000, is as quiet as a motorcycle and as simple to operate as a car.

Although it would not have any road-driving capabilities, it would bring this form of travel within the grasp of a wider section of people. Technology would automate many of the pilot's functions. This Small Aircraft Transportation System (SATS) would divert pressure away from the "hub-and-spoke" model of air travel. Hub-and-spoke refers to the typically US model of passengers being processed through large "hub" airports and then on to secondary flights to "spoke" airports near their final destination.

Source: https://learning.sohu.com/20041008/n222367149.shtml

第3章 词语的翻译技巧

本章中词语是指单词（word）与短语（phrase）。词语是语言中能够独立出现的最小语言单位。尽管我们在翻译中要传递的是语篇的总体意思，但在解读文本时，却常常要从词语入手。然而，我们会发现在绝大多数句子中很难找到英汉完全对应的等值词，因此，词对词的对等翻译，几乎不可能。在多数情况下，英汉词语之间是不对等的。

第1节 英汉词语的对应情况以及词义的选择

辩证唯物主义认为，语言常常是客观世界的反映，是一种社会现象。人们在一种什么样的环境里生活、劳动，就会产生出什么样的语言。由于中国与英语国家在生活环境、经验、风俗习惯、宗教信仰、对客观世界的认识等诸多方面存在巨大的差异，必然在英汉两种语言中有所体现，其中之一就是英汉词语呈现出的复杂的对应情况。英汉词语的词义对应情况大致可归纳为以下四种：

（1）完全对应。

由于人们对客观世界的认识、对自然现象的描述以及社会文化生活存在许多不谋而合之处，即不同文化之间存在文化共核（cultural common core），一些描述自然现象、人体感应、亲缘关系，以及科技的词汇和专有名词在中英文中基本上呈完全对应，例如：风、雨、云、雾、太阳、地球、月亮、大、小、冷、热等都可以在英语中找到对等的表述：wind，rain，cloud，fog，sun，earth，moon，big，small，cold，hot 等。血缘关系如：父亲、母亲、儿子、女儿等在英语中也都有对等的表达：father，mother，son，daughter；科技词汇和专有名词，更容易找到对等的翻译，例如：物理、化学、数学、联合国、欧元、香港等，均可在英语中找到对等的翻译：physics，chemistry，mathematics，the United Nations，Euro，Hong Kong 等。英汉语中单词的对应率最高，其次是词组、成语、谚语。

（2）部分对应。

由于不同的风俗习惯、生活经验等原因，有许多英汉语词语在词义上只是部分对应。它们的意义范围有广狭之分、抽象与具体之分，以及一般与个别、方位与视角、褒义与贬义之分。例如：

uncle(广义)，叔父、伯父、姑父、姨父、舅父（狭义）；
milk 奶（抽象），人奶、牛奶、羊奶（具体）；
film 胶卷（一般），影片（个别）。

再比如，汉语的“借”在英语中就有两个不同的词，向别人借是 borrow，把东西借给别人是 lend，汉语则用同一个“借”字。又如，汉语中“宣传”一词并没有不好的意思，但在英语中 propaganda 就有不好的表达意思。根据《柯林斯伯明翰大学国际语料库》，propaganda—usually used showing disapproval，即，常用于贬义。此外，还有些英汉词语在概念意义上是对应的，但在内涵意义上却是不对应的，例如：

vinegar: 与中文的“醋”在概念意义上是对应的，但有“不高兴、坏脾气”的内涵意义；而“醋”的内涵意义却是“妒忌”。

(3) 不对应。

也就是一些文化空缺词。由于文化差异，英语有些词语所表达的意义在汉语中尚无确切的对应词来表达。它们主要是一些具有文化含义的词语，翻译时多需加注或释义。例如：

hippie 嬉皮士；
breaker 跳布瑞克舞者、跳霹雳舞者；
rigjacker 劫持近海油井设备的人；
eggathon 吃煮硬了的鸡蛋的竞赛；
congressperson 美国国会（尤其是众议院）议员（男议员或女议员）。

(4) 交叉对应。

英语中有许多多义词，其各个意义分别与汉语几个不同的词语对应。如下面各例中的 light 一词就是这样：

light music 轻音乐（light = intended chiefly to entertain）；
light loss 轻微的损失（light = not heavy）；
light heart 轻松的心情（light = relaxed）；
light car 轻便汽车（light = having little weight）；
light step 轻快的步伐（light = gentle）；
light manners 轻浮的举止（light = frivolous）；
light outfit 轻巧的设备（light = handy）
light work 轻松的工作（light = requiring little effort）；
light voice 轻柔的声音（light = soft）

英汉语词语之间复杂的对应情况决定了译者在翻译文本时，必须根据具体语境正确理解词语，选择适当的语义。根据词汇意义的理论原则，如欲判断词语的意义，或有效使用词典，便须牢记有关词义的基本理论原则。首先，任何语境中词汇的正确意义，一般应当是最符合该语境的那个意义。换言之，正确的理解最大限度地依赖于语境，而不是最大限度地依赖于孤立的词语。第二个理论原则是，在任何一个语境里，一个词语只可能有一个意义，而不会有几个意义。第三个原则可看作解释原则，是指词汇字面意义，即无标记意义应被认作正确的意义，除非语境表明具有另一个意义。最后一个也许是最重要的一个原则，即符号总是通过其他符号加以解释的。这些起解释作用的符号分为两个不同的类别：言语符号和非言语符号。例如，对英语单词 chair（椅子）的意义可以利用同一语义场的其他词，如 bench（板凳）、stool（方凳）、throne（宝座）和 pew（教堂靠背长凳）进行解释。但是在市政会议上，Please address the chair 一句中的 chair 则通过非言语的实际环境，标记出它是指人（即会议主席），而不是指家具。

鉴于英汉语中都有一词多类、一词多义现象，在确定英语词义时需要根据句法结构判明一个词在原句中属于哪一词类，然后进一步确定词义。例如，在下面各句中，air 分属不同的词类：

- The rain had cleaned the **air**. 雨后空气新鲜。(名词)
- Don't **air** your troubles too often. 别老是诉苦。(动词)

即使词类不变，一词多义也是很常见的。因此，译者在确定词类的基础上，应根据某一词语使用的专业领域以及上下文来确定其确切的含义。例如，

- Assume that the input voltage from the power supply remains constant.
- Combat power is useless unless it can be brought to bear at the right point, and at the right time.
- The combining power of one element in the compound must equal the combining power of the other element.

在以上各句中 power 一词均用作名词，但应用于不同领域，分别为电学、军事、化学领域，因而其意义分别是："电源、战斗力、化合价"。

任何词语都不是孤立使用的，总是出现在一定的上下文当中，正如俗语所说，词无定义，意随境生。例如，英语中 dry 并非总指"干的"，在一定的上下文中该词具有不同的意义，可以指"不产奶的、不卖酒的、不加渲染的、枯燥无味的、干洗的、无酒的、无甜味的、禁酒的、无水的"等。因此，对于词义的选择，应把词语放在句、段乃至篇章的语言环境中加以考虑，也就是强调语境（语言语境以及文化语境）对词义的限定和选择。

此外，语言不是由词语随意拼凑而成的，词与词之间的搭配是有一定限度的，如某类动词只能跟某种主语或宾语一起出现，某些形容词只能修饰某类名词，某些副词也只能修饰某类形容词、动词或其他副词。而且，英汉两种语言词汇的搭配能力、搭配习惯也是不同的，译者必须具有扎实的语言功底，熟知两种语言的习惯搭配，同时，还要善于利用如《英汉搭配词典》《现代汉语实词搭配词典》等工具来判断和选择合适的搭配词。即使是英语中同一个词译成汉语，也要根据不同情况，采取不同译法，使它合乎汉语搭配习惯。例如 develop 一词，用在不同的地方，译成汉语，就要选择不同的词语来表达：

Many laboratories are **developing** medicines against AIDS.
许多实验室正在研制治疗艾滋病的药物。

Noise may **develop** in a worn engine.
旧发动机会产生噪声。

People who drink and smoke heavily are likely to **develop** cancers of the oral cavity and tongue at a younger age.
大量喝酒抽烟的人可能年纪轻轻就会患上口腔癌和舌癌 。

In **developing** the design, we must consider the feasibility of processing.
进行设计的时候，必须考虑工艺的可行性。

第2节 增译与省译

2.1 增译（Amplification）

英汉两种语言，由于表达方式不尽相同，翻译时要在词量上加以增减。所谓增词译法，就是在原文的基础上添加必要的单词、词组、分句或完整句以使译文明白晓畅。这当然不是无中生有地随意增词，而是增加原文中虽无其词但有其意的一些词，从而使得译文在语法、语言形式上符合译入语习惯，并在文化背景、词语联想方面与原文一致，使得译文与原文在内容和形式等方面对等。增词译法可以划分为语法性增译，语义性增译以及逻辑性增译。

（1）语法性增译。

由于中英文语法上的差异，为了使译文语义明确，畅达通顺，符合汉语的习惯，有时需要增加名词、代词、量词，以及增加英文中表示名词复数、动词时态等方面的助词。

增加量词

英文里没有量词，但中文有量词，因此，根据中文的语法，在一些名词前面需要增译适当的量词。例如：

an elephant 一头大象　a pen 一支钢笔　a martial music 一首军乐曲　a plane 一架飞机 a tree 一棵树　a flower 一朵花，等等。

增加人称代词或做主语的名词

在一些英语的特定句式中，如被动句、形式主语句中需要增加泛指人称代词或者名词做句子的主语。例如：

Weak magnetic fields are known to come from the human body.

我们知道，人体能产生微弱的磁场。

It is estimated that the new synergy between computers and Net technology will have significant influence on the industry of the future.

人们预测，新的电脑和网络技术的结合将会对未来的工业产生巨大的影响。

It's believed that we shall make full use of the sun's energy some day.

我们相信，总有一天我们将能够充分利用太阳能。

With the development of modern electrical engineering, power can be transmitted to wherever it is needed.

随着现代化电气工程的发展，人们可以把电力输送到任何所需要的地方。

The design is considered practical.

大家认为这个设计切实可行。

增加表示复数概念的词语

英文的名词有单复数，而中文的名词无单复数形式。因此，翻译英语的复数名词时，应增译表示名词复数概念的词语，诸如用“们、一些、许多”等词语把英语中表示名词复数的概念译出。例如：

Carbon combines with oxygen to form carbon oxides.

碳同氧化合形成多种氧化碳。

Of visible lights, red light has the longest and violet the shortest wavelength.

在各种可见光中，红光的波长最长，紫光的波长最短。

When the plants died and decayed, they formed layers of organic materials.

植物腐烂后，形成了一层层有机物。

In the 1980s, departments bought their own minicomputers and managers bought their PCs.

20 世纪 80 年代，许多部门都购买了自己的微机，经理们也购买了个人电脑。

There is enough coal to meet the world's needs for centuries to come.

有足够的煤来满足全世界未来几个世纪的需要。

增加表示时间的副词或助词

英文的动词有时态变化，而中文的动词没有时态上的词形变化以及相应的助动词。因此翻译动词时，则应增加相应的时间副词或助词，用来表示不同的时态。例如：

Contemporary natural science **are** now **working** for new important breakthroughs.

当代自然科学正在酝酿着新的重大突破。

Humans **have been dreaming** of copies of themselves for thousands of years.

千百年来，人类一直梦想制造出自己的复制品。

This natural approach to remediating hazardous wastes in the soil, water, and air **is capturing** the attention of government regulators, industries, landowners, and researchers.

这种清除土壤、水以及空气中有害废物的自然方法正日益受到各政府首脑、企业、土地所有者、研究者们的关注。

The high-altitude plane **was** and still **is** a remarkable bird.

高空飞机过去是，而且现在仍然还是一种了不起的飞行器。

（2）语义性增译。

英语中往往大量使用省略结构，但汉译时，一般需要将省略的成分补出，才能使译文句子意思完整，表达通顺。这种按语义需要而增词的译法称作语义性增译。例如：

The changes in matter around us are of two types, physical and chemical.

我们周围的物质变化有两种，物理变化与化学变化。

High voltage is necessary for long transmission line while low voltage for safe use.

远距离输电需要高压，而安全用电则需要低压。

High temperatures and pressures changed the organic materials into coal, petroleum and natural gas.

高温和高压把这些有机物变成了煤、石油和天然气。

Matter can be changed into energy and energy into matter.

物质可以转化为能量，能量也可以转化为物质。

此外，有些英语句子如果直译成汉语，其意思表达语焉不详，因此汉译时必须根据上下文增译适当的词语，例如动词、形容词，或者根据中文的表达习惯增译一些范畴词，表示方法、现象、效应、作用、状态，等等。例如：

Plastic bowls marked microwavable are probably safer than those that aren't.

使用贴有“可用于微波炉烹调”标志的塑料碗也许比使用没有贴这种标志的塑料碗更安全。

The statistics brought out a gender division between hard and soft science: girls tending toward biology, boys tending math and physics.

统计表明，从事硬科学和软科学研究的科学家在性别上存在着差异：女性倾向于从事生物学的研究，而男性则倾向于从事数学和物理学的研究。

Were there no electric pressure in a conductor, the electron flow would not take place in it.

导体内如果没有电压，便不会产生电子流动现象。

In the US, there is evidence that schoolchildren with access to the Internet are starting to watch less TV and spend more time surfing the Net.

在美国，有证据表明，能上因特网的在校学生看电视的时间少了，但在网上冲浪的时间增加了。

Floppy disks have always been cheap to make and relatively easy to copy.

软盘的制作成本一贯比较低，而且相对而言，也比较容易拷贝。

(3) 逻辑性增译。

由于中英文具有不同的句式特点及表达习惯，英文中符合逻辑的，直译成中文不一定符合中文习惯，这时应增译适当的表示原因、条件、目的、结果、让步、假设等逻辑关系的连接词，以使译文通顺流畅。例如：

Ice and water consist of the same substance in different forms.

冰和水虽然由相同的物质构成，但形态不同。

PCs are not the most stable, and with video editing you have to be extra careful.

个人电脑一般并不太稳定，所以，进行视频编辑时需要加倍小心。

In the absence of sufficient knowledge confusion exists in classifying a specimen as a butterfly or a moth.

如果没有足够的知识，就分不清标本是属于蝶类还是蛾类。

2.2 省译（Omission）

省译指把原文中需要而译文中不需要的词、词组等在翻译过程中加以省略，使译文更加通顺。但需要注意的是，省译并不是说译者可以随心所欲任意删减原文的内容，而是指省去原文有但译入语表达不需要，且减去也不影响表达原文意思的词语。省译的操作多出于译入语的语法和习惯表达的需要而进行。省译可以划分为语法性省译、引导词的省译与语义性省译。

(1) 语法性省译。

语法性省译又称虚词的省译，指英译汉时在不损害准确传达原文内容的前提下，根据具体情况将英文中的冠词、连词、介词等虚词略去，使译文练达晓畅。

省译冠词

英语中有冠词，而汉语中没有冠词，因此，当冠词用于名词之前，表示某一类别的人或者事物时，汉译时常常可以省去不译。例如：

Much of the sunlight is absorbed by the oceans.

大部分阳光被海洋吸收。

A square has four equal sides.

正方形四边相等。

Such an engine is called an internal combustion engine.

这种发动机称作内燃机。

省译介词和连词

由于中英文句法结构的差异，英语中大量使用的介词与连词也常常予以省译，以使译文更加简洁凝练。例如：

The symptoms **of** hepatitis are similar no matter which virus is involved.

不管是由何种病毒引起的，肝炎症状均大同小异。

The fact that the concrete is slow **in** setting is no sign that it is **of** poor quality.

混凝土凝固得慢，并不表示其质量差。

Check the circuit **before** you begin the experiment.

检查好线路后再开始做实验。

The volume of a given weight of gas varies directly with the absolute temperature, **provided** the pressure does not change.

压强不变，一定重量的气体体积与绝对温度成正比。

The inner and outer rings are graded **and** stored according to size.

内外圈是根据尺寸分类存放的。

(2) 英语特殊句式结构中引导词的省译。

There + be (seem, appear, exist, happen) 句型是英语特有的句型结构，汉译时可不必强行翻译出来，即不把“有”翻译出来；另外，在 it 引导的形式主语句或者形式宾语句中，it 也经常予以省译。例如：

There is no cloud in the sky.

天空万里无云。

There exist neither perfect insulators nor perfect conductors.

既没有绝对的绝缘体，也没有绝对的导体。

It was not until the middle of the 19th century that the blast furnace came into use.

直到19世纪中叶，高炉才投入使用。

Scientists have proved it to be true that the heat got from coal and oil comes originally from the sun.

科学家们已证实，从煤和石油中获得的热能都源于太阳。

(3) 语义性省译。

语义性省略又称实词的省译，指译者可以根据译入语的表达习惯和方式省略原文中的名词、动词、代词、形容词，以使译文的表达更通顺流畅，言简意赅。

名词的省译

英文同位语结构中的名词，以及一些固定短语中重复表达的名词在翻译中可以省略，以使译文简洁流畅。例如：

Apart from the **fact** that it may become hot, no apparent change occurs in a metal when it is carrying a current.

分析：

fact 的同位语从句具体说明了 fact 的实质，因此，fact 无须再译出来。

译文：

金属有电流通过时，除了可能变热外不会产生显著的变化。

For the **purpose** of discussion, let's neglect friction.

分析：

"for"一词已经暗含了"目的"意义，因此，purpose 无须再译出。

译文：

为了便于讨论，我们将摩擦力忽略不计。

动词的省译

英语的句子必须由动词作谓语，而汉语则不然，句子中除了动词作谓语，形容词、名词或词组都可以作谓语。所以，原句中的动词汉译时可以酌情予以省译。

例如：

This laser beam **covers** a very narrow range of frequencies.

这种激光束的频率范围很窄。

When the design is complete, the system may then be used to produce detailed engineering drawings.

设计完成时，该系统就可用于生成详细的工程图。

This chapter **provides** a brief review of the theory of modification.

分析：

provide 为与名词化结构配合使用的动词，并无实际意义，在翻译中也可予以省略。

译文：

本章简单回顾一下诱发变异理论。

代词的省译

英语中大量使用各种形式的代词，这些代词多数可以予以省译。例如：有泛指含义的人称代词“we, you”和不定代词“one”；形式主语句及形式宾语句中的“it”；句子中曾出现过的某一名词的人称代词或指示代词；用作定语的物主代词。例如：

You will see to **it** that the engine doesn't go out of order.
注意别让发动机出故障。(You 和 it 均省译)

The finished products must be sampled to check **their** quality before **they** leave the factory.
成品出厂前必须抽样进行质量检查。

The waste gases are harmful to us and we should by all means remove **them**.
废气对我们有害，应尽力加以排除。

形容词的省译

英语中一些形容词与其修饰的名词语义重复，译出来反而会引起歧义，应该予以省译。例如：

This treatment did not produce any **harmful** side effect.

分析：

side effect: 副作用，指的就是不好的作用，与 harmful 语义重复，因此将 harmful 省译。

译文：

这种治疗方法不会产生任何副作用。

第 3 节　转性译法（Conversion）

英汉语遣词造句的习惯不同，英汉互译时，原文中的某些词语在译文中要转换词类，比如名词转为动词或形容词，形容词转为名词或副词，动词转为名词等。这种翻译方法就是转性译法，又称词类转换，是英汉互译时最常用的一种变通手法。

3.1　转译成汉语的动词

英语和汉语比较起来，汉语中动词用得比较多，这是一个特点。往往在英语句子中只用一个谓语动词，而在汉语中却可以几个动词或动词性结构连用。因此，英语中不少词类（尤其是名词、形容词、副词）在汉译时往往可以转译成动词。

（1）名词转译成动词。

英语中大量具有动作意义的名词和由动词派生出来的名词往往可以译成汉语的动词。例如：

The **application** of electronic computers makes for a tremendous rise in labor productivity.
使用电子计算机可以大大提高劳动生产率。

Computers can provide **analyses** of every operation in a factory.
电子计算机可以分析工厂的每道工序。

We must place stress on the **prevention** of diseases.
我们应以预防疾病为主。

(2) 形容词转译成动词。

英语中某些表示情态、感受、感觉、信念等意义，在句中作表语的形容词以及同介词搭配构成句子表语或定语的形容词，通常可译成汉语的动词。例如：

The amount of work is **dependent on** the applied force and the distance the body is moved.
功的大小取决于所施加的力与物体所移动的距离。

Kerosene is not so **volatile** as gasoline.
煤油不像汽油那样容易挥发。

Heat is a form of energy into which all other forms are **convertible**.
热能是能量的一种形式，其他一切形式的能量都能转化为热能。

(3) 副词转译成动词。

英语中有些作表语的副词或复合宾语中的副词，往往可译成汉语中的动词。例如：

The test was not **over** yet.
试验还没结束。

Open the valve to let air **in**.
打开阀门，让空气进入。

An exhibition of new products is **on** there.
那里正在举办新产品展览会。

The experiment in chemistry was ten minutes **behind**.
这个化学试验延误了10钟。

The electric current flows through the circuit with the switch **on**.
如果接通开关，电流就流过线路。

(4) 介词转译成动词。

英语中的介词或介词短语在许多情况下可以译成汉语的动词。例如：

The volume of a gas becomes smaller when the pressure **upon** it is increased.
当作用在气体上的压力增大时，气体的体积就缩小。

The letter E is commonly used **for** electromotive force.
通常用字母E表示电动势。

No body at rest can be set in motion without a force being acted **upon** it.
没有受到力的作用，静止的物体不可能开始运动。

It is clear that numerical control is the operation of machine tools **by** numbers.
很显然，数控是指机床采用数字来操纵。

3.2 转译成汉语的名词

(1) 动词转译成名词。

英语中有些动词在翻译成汉语时很难找到相应的动词，这时可将其转译成汉语相关意义的名词。例如：

Neutrons **act** differently from protons.
中子的作用不同于质子。

Gases **differ** from solids in that the former have greater compressibility than the latter.
气体和固体的区别在于前者比后者有更大的可压缩性。

The electronic computer is chiefly **characterized** by its accurate and rapid computations.
电子计算机的主要特点是计算准确而且速度快。

Mercury **weighs** about thirteen times as much as water.
水银的重量约为水的13倍。

(2) 形容词转译成名词。

英语中有些形容词加上定冠词the表示某一类人或物，汉译时常译成名词。另外，英语中有些表示事物特征的形容词作表语时，往往也可在其后加上“性”“度”“体”等词，译成名词。例如：

Both the compounds are acids, **the former** is strong and **the latter** weak.
这两种化合物都是酸，前者是强酸，后者是弱酸。

The atmosphere is only about eleven kilometers **thick**.
大气的厚度约为11千米。

People are so much more flexible and **inventive** than robots.
人类比机器人更灵活而且更富有创造性。

In fission processes the fission fragments are very **radioactive**.
在裂变过程中，裂变碎片的放射性很强。

The more carbon the steel contains, the **harder** and **stronger** it is.
钢的含碳量越高，强度和硬度就越大。

(3) 代词转译成名词。

英语中的代词译成汉语，有时为了使译文规范、意义准确，须转换为其所指代的名词。例如：

We need frequencies even higher than **those** we call very high frequency.
我们所需要的频率，甚至比我们称作高频率的频率还要高。

Air density decreases as the temperature goes up and **it** increases when **it** gets colder.
气温升高时，空气密度就减小；气温变冷时，空气密度就增大。

Though we cannot see **it**, there is air all around us.
虽然我们看不见空气，可我们周围到处都有空气。

The volume of the sun is much larger than **that** of the earth.
太阳的体积比地球的体积大得多。

(4) 副词转译成名词。

英语中有些副词，在句子中用作状语，译成汉语时可根据具体情况转换成汉语相关意义的名词。例如：

Such magnetism, because it is **electrically** produced, is called electromagnetism.

由于这种磁性产生于电，所以称为电磁。

India has the software skills and thousands of software developers who are English-speaking and **technically** proficient.

印度拥有软件设计方面的技术，而且拥有成千上万能说英语、精通技术的软件工程师。

The computer is shown **schematically** on this page.

这一页是计算机的简图。

It was not until early 40's that chemists began to use the technique **analytically**.

直到40年代初，化学家们才开始将这种技术用于分析工作。

3.3 转译成汉语的形容词

（1）副词转译成形容词。

Nitric acid is an **extremely** reactive agent.

硝酸是一种强烈的反应剂。

The wide application of electronic computers affects **tremendously** the development of science and technology.

电子计算机的广泛应用对科学技术的发展有极大的影响。

Amphibians and birds do differ **significantly**.

两栖类和鸟类之间有显著的差别。

This new electronic computer is **chiefly** characterized by its simplicity of structure.

这种新型电子计算机的主要特点是结构简单。

（2）名词转译成形容词。

一些作表语的名词或由形容词派生的名词，翻译时往往可以根据语义转译为汉语的形容词。例如：

Electronic computers and microprocessors are of great **importance** to us.

电子计算机和微处理器对我们来说都十分重要。

The experiment was quite a **success**.

这次试验很成功。

Most teenagers feel no **difficulty** in learning and operating computers.

绝大多数青少年在学习和操作电脑方面并不觉得困难。

In certain cases friction is an absolute **necessity**.

在一定情况下，摩擦是绝对有必要的。

3.4 转译成汉语的副词

（1）形容词转译成副词。

一些修饰名词的形容词，由于名词转成动词，因而相应地转译成副词；有时为了使译文符合汉语表达习惯，也要把英语形容词转译为汉语副词。例如：

With **slight** repairs the old type of engine can be used.
只要稍微修理一下，这台老式发动机就可使用了。

The engine has given a constantly **good** performance.
这台发动机一直运转良好。

When **precise** timing is necessary, care must be taken not to interrupt the timing wave.
如果需要精确定时，必须注意不要中断定时波。

A helicopter is **free** to go almost anywhere.
直升机几乎可以自由地飞到任何地方。

The **same** principles of low internal resistance also apply to milliammeters.
低内阻原理也同样适用于毫安表。

（2）其他词类转换成副词。

The added device will ensure **accessibility** for part loading and unloading.
增添这种装置将保证工件便于装卸。(名词转译为副词)

They reported they had **succeeded** in turning normal human cells into cancerous one.
他们报告说已成功地把正常的人类细胞转化为癌细胞。(动词转译为副词)

Microscope **continues** to be a very important tool in science today.
今天，显微镜依旧是科学研究中一个很重要的工具。(动词转译为副词)

第4节　词义的引申（Semantic Extension）

英译汉时，有时会遇到某些词在词典中找不到适当的词义，如果生搬硬套词典中的释义，译文就不能确切地表达原文的意思，甚至会造成误译。这时就应结合上下文和逻辑关系，根据汉语的表达习惯，引申词义。所谓词义引申，就是指对某些英语词语的含义加以扩展和变通，使其更准确地表达原文词语所表达的特定意思。词义的引申主要包括以下三个方面的内容：

4.1　专业化引申（Extension According to Specialty）

英语科技文章所使用的词汇有些借自日常生活用语，但在科技语境中已被赋予不同于日常语境的专业化语义。因此，翻译时应基于其基本意义，根据所涉及专业，引申出其专业化语义。例如：

His mother died of **difficult labor**.
他母亲死于难产。(不译作“困难的劳动”)

The major problem in manufacturing is the control of contamination and **foreign material**.
制造时的一个主要问题是控制污染和杂质。(不译作“外界物质”)

The **running** of such automated establishments remains only a matter of **reading** various meters mounted on panels.

管理这种自动化工厂只不过要求查看一下控制台上的各种仪表而已。（不译作“奔跑”和“阅读”）

More weight must be placed on the **past history** of patients.

必须更加重视患者的病史。（不译作“过去的历史”）

4.2 具体化或形象化引申（Extension by Concretization）

英语科技文章中有些字面语义颇为笼统或抽象的词语，若按字面译出，要么不符合汉语表达习惯，要么就难以准确地传达原文所表达的意思，在这种情况下，就应该根据特定的语境，用比较具体或形象化的汉语词语对英文词语所表达的词义加以引申。例如：

Other **things** being equal, copper heats up faster than iron.

其他条件相同时，铜比铁热得快。（不译作“事情”）

There are many **things** that should be considered in determining cutting speed.

在测定切削速度时，应当考虑许多因素。（不译作“事情”）

Her soft features effectively hid her strong mechanical **foundation**.

她柔媚的容貌巧妙地掩饰了她那强有力的机械躯体。（不译作“基础”）

Steel and cast iron also differ in **carbon**.

钢和铸铁的含碳量也不相同。（不译作“碳”）

4.3 概括化或抽象化引申（Extension by Abstraction）

英语科技文章中有些词语的字面意思比较具体或形象，但若直译成汉语，有时则显得牵强，不符合汉语的表达习惯，使人感到费解。在这种情况下，就应用含义较为概括或抽象的词语对英文词语所表达的词义加以引申。例如：

Americans every year **swallow** 15,000 tons of aspirin, one of the safest and most effective drugs invented by man.

阿司匹林是人类发明的最安全、最有效的药物，美国每年要消耗15 000吨。（不译作“吞咽”）

Alloys belong to a **half-way house** between mixtures and compounds.

合金是介于混合物和化合物之间的一种中间结构。（不译作“两地间中途歇脚的客栈”）

In 24 hours the heart in a human body **receives and pumps** out some 1,000 quarts of blood.

24小时内，通过人的心脏循环的血液约为1 000夸脱。（不译作“接受或泵出”）

Superconductivity technology is now **in its infancy**.

超导技术正处于发展初期。（不译作“处于婴儿期”）

第5节 重复译法（Repetition）

英语行文往往比较忌讳重复，因此常会出现一个动词后面接几个宾语或者表语的现象，或者大量使用代词以避免重复使用某个名词的现象。但中文行文方式有所不同，在翻译时，有时为了使行文更明确，表达更生动，或者为了强调，我们往往要将一些关键性的词加以重复。重复译法不同于增译法，前者是在译文中重复使用已出现过的词语，而后者则是在译文中根据需要增补新的词语。

为了使行文表达更明确而重复

We should learn how to analyze and solve problems.

分析：

英文原文是两个动词带一个宾语，翻译时应重复宾语使行文明确。

译文：

我们应学会如何分析问题和解决问题。

I had experienced oxygen and/or engine **trouble**.

分析：

原文中宾语 trouble 有两个修饰语，为了使意思明确，翻译时将宾语重复。

译文：

我曾碰到过，不是氧气设备出故障，就是引擎出故障，或两者都出故障的情况。

The doctor will get more **practice** out of me than out of one hundred ordinary patients.

分析：

英语中常重复使用介词，而将第二个、第三个介词前的名词省略，翻译时则往往要把此名词重复。

译文：

医生从我身上得到的实践会比从一百个普通病人身上得到的实践还多。

The three most important effects of an electric current are **heating, magnetic and chemical effects**.

分析：

英语句子中几个形容词共同修饰某一个名词时，被修饰的名词在译文中常常需要重复。

译文：

电流三种最重要的效应是热效应、磁效应和化学效应。

Matter changes not only in state but also in volume.

分析：

英语中一些并列连词诸如：not only…but also，as well as，but，not…but，either…or…，both…and…，neither…nor…连接两个并列成分，且只有一个谓语动词时，英译汉

时可以重复翻译出该动词，以使译文表达明确。

译文：

不仅物质的状态会变化，体积也会变化。

第6节 练习

一、思考题

1. 英汉语词语之间的对应情况有哪几种？在确定某个词语的意义时，译者应该考虑哪几个因素？

2. 什么是增译法，什么是省译法？分别有哪几种类型的增译与省译？

3. 什么是转性译法？英语中什么词类可以分别转译成汉语的动词、名词、形容词与副词？

4. 什么是词义的引申？分别有哪几种类型的引申？

5. 什么是重复译法？与增译法有何不同？

二、翻译实践题

1. 将下面的科技术语译为中文

idle capacity idle coil idle current idle frequency idle wheel idle roll

idle space idle contact idle motion idle stroke

solid angle solid bearing solid body solid line solid borer solid gold

solid crankshaft solid circuit solid measure solid injection

dry fire dry cow dry shampoo dry wine dry farming

2. 增译、省译翻译练习

（1）Things in the universe are changing all the time.

（2）Air is a mixture of gases.

（3）By the turn of the 19th century geologist had found that rock layers occurred in a definite order.

（4）The earth's population is doubling, the environment is being damaged.

（5）This transfer continues until a uniform temperature is reached, at which point no further energy transfer is possible.

（6）Potassium and sodium are seldom met in their natural state.

（7）Patients may have to take the new medicine for the rest of their lives, and the expense and complexity of the regimen keep them out of reach for the 9 out of 10 patients who live in developing nations.

（8）Matter can be changed into energy and energy into matter.

（9）Air pressure decreases with altitude.

（10）During an El Nino the pressures over Australia, Indonesia and the Philippines are higher than normal, which results in dry conditions or even droughts.

（11） The direction of a force can be represented by an arrow.

（12） The mechanical energy can be changed back into electrical energy by means of a generator or dynamo.

（13） In human society activity in production develops step by step from a lower to a higher level.

（14） Very significant advances in lasers have been made in the past several years that open the door to more rapid progress in optical communications systems.

（15） Obviously, since this radiation is invisible, it is of no use to us for illumination purpose.

（16） Apart from the fact that it may become hot, no apparent change occurs in a metal when it is carrying a current.

（17） If there is little cooling water available, a diesel engine makes an excellent prime mover for generation below 10,000 kVA.

（18） There was a rise in temperature, and as a result, the component failed.

（19） This critical deficiency was itself due to three factors which interacted with each other over the entire period.

（20） If the surface is smooth, reflection is regular; if the surface is rough, the reflection is diffuse.

3. 转性译法翻译练习

（1） All of this proves that we must have a profound study of properties of metals.

（2） This workpiece is not more elastic than that one.

（3） Rapid evaporation at the heating surface tends to make the steam wet.

（4） Some employers are keen to encourage employees to go green, and switch to smaller cars.

（5） The wide application of electronic machines in scientific work, in designing and in economic calculations will free man from the labor of complicated computations.

（6） With slight modifications each type can be used for all three systems.

（7） Continual smoking is harmful to health.

（8） Below 4℃ water is continuous expansion instead of continuous contraction.

（9） The nuclear power system designed in China is of great precision.

（10） While tallness is evidently a hereditary characteristic, any individual's actual height depends on the interaction between their genes and the environment.

（11） Robotics is so closely associated with cybernetics that it is sometimes mistakenly considered to be synonymous.

（12） The engineer had prepared meticulously for his design.

（13） Gases conduct best at low pressures.

（14） The computer exhibition impressed us deeply.

（15） The universal lathe is most widely used and plays an important part in industry.

（16） The experiment was quite a success.

（17） Oxygen is one of the important elements in the physical world, it is very active chemically.

（18） The quality of the operating system determines how useful the computer is.

（19）The image must be dimensionally correct.

（20）The only naturally occurring substance used as fuel for nuclear power is U-235.

4. 词义的引申翻译练习

（1）The year 2000 problem brings the largest companies in the world to their knees.

（2）Industrialization and environmental degradation seem to go hand in hand.

（3）In the microbial transformation of contaminants, organisms can either "eat" the toxins or break them down in the process of consuming other substances.

（4）The research performed on smokers at rest indicates that smokers burn more calories than nonsmokers.

（5）The plan for launching the man-made satellite still lies on the table.

（6）The furnace eats up fuel at the rate of three tons per hour.

（7）The thermometer rises or falls according as the air is hot or cold.

（8）Australian researchers have developed an implantable hearing prosthesis that inserts 22 electrodes into the inner ear and is connected to a pocket speech processor.

（9）For the average consumer, voice-activated devices are a convenience; for the elderly and handicapped, they may become indispensable for a wide variety of chores in the home.

（10）In general, the design procedure is not straightforward and will require trial and error.

（11）A large segment of mankind turns to untrammeled nature as a last refuge from encroaching technology.

（12）The bacterial pneumonia may complicate influenza at both extremes of age.

（13）The doctor may insert thin needles into the skin of a patient at key points along meridians.

（14）Everything is ready, so I call the morning paper up on the screen.

（15）This kind of wood works easily.

（16）This medicine acts well on the heart.

（17）We shall develop the aircraft industry in a big way.

（18）Some employers are keen to encourage employees to go green, and switch to smaller cars.

（19）The doctors were operating on the patient, but the patient's nerve was continually on the stretch.

（20）Copper wire is flexible.

5. 重复译法翻译练习

（1）PCs gave the world a whole new way to work, play and communicate.

（2）Ice is not as dense as water and it therefore floats.

（3）The sun is regarded as the chief source of heat and light.

（4）The equations below are derived from those above.

（5）The most common acceleration is that of free falling bodies.

（6）Energy can neither be created, nor destroyed, although its form can be changed.

（7）Continuous data can be expressed either in fractions or whole numbers.

(8) Matter changes not only in state but also in volume.

(9) Radio waves are similar to light waves except that their wave length is much greater.

(10) The electrolytic process for producing hydrogen is not so efficient as the thermo chemical process.

(11) The force due to the motion of molecules tends to keep them apart.

三、拓展阅读和翻译练习

阅读以下科技语篇，并翻译划线句子以及加黑段落。

In Cities, Flooding and Rainfall Extremes to Rise as Climate Changes

Cities face harsher, more concentrated rainfall as climate change not only intensifies storms, but draws them into narrower bands of more intense downpours, UNSW engineers have found. This has major implications for existing storm-water infrastructure, particularly in large cities, which face higher risks of flash flooding.

In the latest issue of Geophysical Research Letters, doctoral student Conrad Wasko and Professor Ashish Sharma of School of Civil and Environmental Engineering at the University of New South Wales show the first evidence of storm intensification triggering more destructive storm patterns.

"As warming proceeds, storms are shrinking in space and in time," said Wasko, lead author of the paper. "They are becoming more concentrated over a smaller area, and the rainfall is coming down more plentifully and with more intensity over a shorter period of time. When the storm shrinks to that extent, you have a huge amount of rain coming down over a smaller area."

Wasko and Sharma, working with collaborators at the University of Adelaide, analyzed data from 1,300 rain gauges and 1,700 temperature stations across Australia to see how air temperature affected the intensity and spatial organization of storms.

They found that atmospheric moisture was more concentrated near the storm's center in warm storms than in cooler ones, resulting in more intense peak rainfalls in those areas. The storms were clearly shrinking in space, irrespective of the amount of rain that fell.

Although the data is sourced from Australia, this has global implications, said Sharma. 1. "Australia is a continent that spans almost all the climate zones in the world—Mediterranean, tropical, temperate, subtropical—everything except the Arctic or Antarctic." So, the results hold a lot of value—we are finding the pattern repeating itself over and over, happening around Australia and around the world.

"Look at the incidents of flooding in Mumbai or in Bangkok last year—you see urban streets full of water," he added. "You see it now in Jakarta and Rome and many parts of Canada. That's because the storm-water infrastructure cannot handle the rain, and part of the reason there's more rain is the increase in global temperatures."

2. Most urban centers have older storm-water infrastructure designed to handle rainfall patterns of the past, which are no longer sufficient. "The increase is especially noticeable in urban centers, where there is less soil, unlike rural areas, to act as a dampener," said Sharma. "So, there is often

nowhere else for the water to go, and the drainage capacity is overwhelmed. So, the incidence of flooding is going to rise as temperatures go higher."

Wasko, lead author of the paper, said scientists have long suspected that the intensity of rainfall would be boosted by climate change, as the warming air raises the carrying capacity of moisture. 3. But while extreme rainfall has been rising, little was known about the mechanisms causing it. The latest study shows that storms are changing in spatial terms.

It follows a study by the same authors in Nature Geoscience in June 2015 showing that storms were also changing their "temporal pattern" —that is, getting shorter in time, thereby intensifying. 4. When it comes to flash flooding, the amount of rain that falls over a period of time is much more important than the total volume of rainfall that a given storm delivers. This study was the first to show that climate change was disrupting the temporal rainfall patterns within storms themselves.

If both spatial and temporal changes in storms continue, as they are likely to do as the world warms, there will be more destructive flooding across the world's major urban centers.

5. In their Nature Geoscience paper, the duo calculated that floods in some parts of Australia would likely increase by 40%, especially in warmer places like Darwin. "If you add the spatial pattern from this latest paper, you will probably increase this 40% number to maybe 60%," said Sharma.

Earlier this year, a pivotal framework for infrastructure maintained by the Institution of Engineers, the Australian Rainfall and Runoff national guidelines were updated for the first time since 1987, a process that took three years. It's now clear, said Sharma, that these will need to be adjusted, as the safety and sustainability of Australian infrastructure adapts to a warming climate.

And there are still unknowns to contend with, he added.

"When we say that the storms are shrinking in space and shrinking in time, and we say floods will increase, we are making an assumption that the volume of water coming down is not changing," said Sharma. "That assumption is very conservative, because you would expect the air to hold more moisture. If you factor that in as well, there'll be even more rainfall, and more floods."

Source: https://www.sciencedaily.com

reading	阅读；读物	读数
mouse	老鼠	鼠标（器）
conduct	进行；指挥	传导
compatible	协调的	兼容的

B组（特点：科技词义多样，因专业而异），例如：

base 普通词义：基础；底部
化工：碱；盐基
药学：主剂
纺织：固色剂；媒染剂
数学：底边，基线；基数
建筑：垫板，脚板
军事：基地

dog 普通词义：狗
机械：车床的夹头；止动器
电子：无线电测向器
船舶：水密门夹扣
天文：大犬座；小犬座
气象：雾虹；（预示有雨的）小雨云

carrier 普通词义：搬运工人
计算机：媒体
集成电路：载体
无线电：载波；载波电路
半导体：载流子
机械：托架
航空：运输机
航天：运载火箭
交通：搬运汽车
海军：航空母舰

rectify 普通词义：纠正；整顿
数学：（曲线）求长；从切
化工：精馏；提纯
机械：调整；拨准（仪表）
焊接：直流探伤
电气电子：整流；检波

element 普通词义：因素；要素
机械：零件；构件；部件
无线电：元件；器件
气象：自然力；风雨
植物：原种
动物：生存环境；活动范围
计算机：单元；基元
语法：音素
数学：元；素；诸元
电学：电池；电极；电阻丝
化学：元素；成分
通信：电码
军事：部队；机组；小分队

在科技、科普翻译中，通用技术词语的翻译是一大难点：一是词汇量大，二是其专业词义较难把握，尤其是B组词类。翻译时，译者应仔细查阅科技词典，必要时应请教专业技术人员核实。

（3）普通英语词汇。

普通英语词汇又称半科技词汇（Semi-ST Words），或者半术语。科技英语中的专业词汇，有少量是利用日常的英语词汇语义外延，即赋予它们以新的词义而构成的。这类词与功能词构成科技、科普文的基础词汇，其意义和用法类似于普通英语。而科技词汇的语义外延是根据所指的关系，而不是根据能指的关系来确定。例如：feed（喂养）转译为“供水、输送、电源”等；head（头）转译为“源头、机关、弹头、磁头”等；power（力，权力）转译为“电能、电力、分辨率、幂”；base（基础）在化学领域中转译为“碱”，在医学领域转译为“主药”；dog（狗）在建筑学上指“铆件”，在机械领域译为“齿轮栓”；bowl（碗）用在地质学语域中转译为“盆地”或“斜向中心的地面”，在机械和铸造专业中转译为“碗形结构”。在整个专业词汇中，这种词汇所占的比例不大，但它们往往代表着一些非常重要的基础性概念。

1.2 科技词汇的形成方式及词义的确定

对于科技词汇，如果是首译，根据不同的翻译对象及词汇的类型，译者可以采用的翻译方法有直译、音译、意译、音译加注、借译等方法。而经过多年中外各个领域的交流，对于大多数译者而言，科技词汇的翻译与其说是翻译，不如说是查阅术语词典与选择词义的问题。

纯专业词汇的意义精确且单一，只要确定了专业领域，大多借助术语词典就可以解决。

然而，由于当今科学技术的发展日新月异，可能会有很多新专业术语未能及时编入词典，这时译者就需要基于科技常识、科技词汇的形成及构成方式来翻译，并应请教专业技术人员核实。

科技词汇的形成可分为三种方式，分别是借用、构词法生成和专有名词命名法。

（1）借用。

借用可分为语际借用、语内借用与学科间的借用。语际借用的词汇以法语、拉丁语、古希腊语为主，诸如，来自法语的 machine（机器）、menu（菜单），来自拉丁语的 nucleus（核）、formula（公式），来自古希腊语的 analysis（分析）、criterion（标准）等。语内借用则是半术语产生的重要途径，这样，普通词汇被赋予特定意义后就成为某个专业领域的术语。例如，在计算机领域 server 的意义是“服务器”，其普通意义是“服伺者”；worm 是“蠕虫病毒”，本义是指“虫子”；package 是“软件包”，本义是指“包裹”。而学科间的借用是通用技术词语产生的重要途径，指某些学科或者专业往往会从别的学科专业借用一些概念或者词汇，从而产生本领域的新术语。例如，morphology（形态学）一词，不仅应用于生物学、地质学，也应用于语言学、社会学；而 input（输入）、output（输出）则借自电学。

（2）构词法生成。

构词法生成的一个方式是缀合法，又称派生法（derivation），指利用前缀（prefix）或者后缀（suffix）与词干（stem）一起构成单词。前缀法构词大多改变词义，而单词的词性不变，例如 anti-，表示“反对”，加在词干 body 之前，变成 antibody 其意思就成了“抗体”；non-表示“否定、相反”，nonnuclear 的意思就是“非核的”；pre-表示“预先”，prefabricate 则表示“预先制造”。后缀法构词大多与原词意思相近，但词性发生改变。英语后缀包括名词后缀、动词后缀、形容词后缀，以及副词后缀。例如，名词后缀-ity、-ness、-cy 分别加在三个形容词 elastic、accurate、frequent 之后生成新词则变成 elasticity（弹性）、accurateness（精确度）和 frequency（频率）。因此，译者应该熟悉词缀的意义，以及词汇的构成方式，这对词汇的理解及翻译是大有益处的。

构词法生成新词的另一个方式是复合法，指两个或两个以上的词合成一个新词的构词方法，大多构成名词、动词、形容词，以名词居多。复合词有三种书写形式：1）不带连字符的复合词式，如 hovercraft（气垫船）、crankshaft（曲轴）、airtight（密闭的）、airdrop（空投）；2）带连字符的复合词，如 liquid-hammering（液击）、sugar-free（无糖的）、energy-consuming（消耗能量的）、dry-clean（干洗）、mass-produce（大批量生产）；3）分写式复合词，指各组成单词分开书写，如 satellite antimissile observation system（卫星反导弹观测系统）、flight test（飞行试验）、dog house（高频高压电源屏蔽罩），等等。复合词大多可以直译，但不可一味望文生义，确定其汉语译名前一定要查阅相关工具书，如：machine tool 并非“机器工具”，而是“机床”；bull's eye 不是“牛眼”，而是“靶心”；同样，上文中的 dog house 也绝非“狗窝”，而是“高频高压电源屏蔽罩”。遇到这种字面意思与语境有严重出入的词语，一定要查阅工具书，根据语境、语篇的主题来选择恰当的词义。

此外，拼缀法（blending）也是利用构词法生成术语的一种方式，是指将两个词各取其中的一部分（或者只截取其中一个词的一部分，另一个词则保持不变），以此合成新词的方法。一般情况下，新词的前半部分表示属性，后半部分表主体，例如：smog（smoke + fog，烟雾）、heliport（helicopter + airport，直升机场）、gravisphere（gravity + sphere，引力范围）、biorhythm（biological + rhythm，生物节律）。这一类词大多是名词，通常可以直译。

（3）专有名词命名法。

英语中有许多科技领域的新生事物或新发现是以专有名词命名的，即以人名、地名、品牌名给相关事物命名。例如，电流单位安培就是以电磁现象的主要发现者——安德烈·玛丽安培的姓氏命名的。还有一些来自地名与品牌名称的术语。这些词汇大多属于专业科技词汇，其意义是确定的，且使用范围狭窄，通过查阅词典即可解决。

1.3 通用科技词语及半科技词语的翻译

上文中提到一些通用科技词语广泛应用于不同专业，并且在不同专业中具有不同的意义，因此，首先要根据语境确定某个词语涉及的专业领域，通过同一语义场内的类别参照物、语法线索等方法确定词义，以及合适的译名。例如：

- Assume that the input voltage from the power supply remains constant.
- Combat power is useless unless it can be brought into full play at the right point, and at the right time.
- The combining power of one element in the compound must equal the combining power of the other element.

分析：

这三句话中 power 一词分别用于电学、军事以及化学领域，因而具有不同的专业意义，分别是“电源”“战斗力”及“化合价”的意思。

译文：

- 假定由电源输入的电压保持不变。
- 战斗力如果不能在适当的时间、地点充分发挥作用，就等于零。
- 化合物中一种元素的化合价必须等于另一种元素的化合价。

PAS Peugeot Citroën today unveils two demonstrators featuring diesel-electric hybrid powertrain, the Peugeot 307 and the Citroën C4 Hybride HDi.

分析：

PAS Peugeot Citroën 为法国标致雪铁龙集团，是法国著名的汽车制造公司，本句话大意是：公司推出两款新型车，由此可判断本句话涉及汽车行业，句子中 diesel-electric 与 powertrain, two demonstrators 与 the Peugeot 307 and the Citroën C4 Hybride HDi 互为参照，显然，复合词 powertrain 不可能是字面意思的组合“动力火车”的意思，根据上下文，应该是指车上的动力系统，那么 diesel-electric 与此相应，指的是采用柴油和充电电池的混合动力；two demonstrators 与 the Peugeot 307 and the Citroën C4 Hybride HDi 相呼应，则不能理解为“示范者”，而应结合语境理解为“展品车”。则 two demonstrators 译为“两款展品车”；diesel-electric 译为“柴油－电力”；powertrain 则译为“动力系统”。

译文：

标致雪铁龙集团（PAS Peugeot Citroën）今天推出了两款配有柴油－电力混合动力系统的展品车，即标致 307 型和雪铁龙 C4 Hybride HDi 型。

By early 1971 the company's motive power roster included about 50 active steam engines, only three mainline diesels, and several diesel switchers.

分析：

本句话中涉及铁路系统，diesel 一般情况下指的是柴油，但如果做柴油讲的话，应该是不可数名词，而此处却用了复数形式，显然在此句中应该用作可数名词，在句中它与 steam engines 与 diesel switchers 显然是互为参照，应属于同类物质，那么由此推断，此处 diesels 必然指的是 diesels engines，则根据我国铁路部门的习惯用语，可译为“内燃机车”。

译文：

截止到 1971 年年初，该公司仍有约 50 台蒸汽机车在运行，而干线内燃机车则只有 3 台，内燃调车机车也只有几台。

The wireless sensor network is a useful tool for managing the automatic controls inside the green house.

分析：

controls 在此用作可数名词，其所指显然已经具体化，不可能是表示抽象的动作意义，因此，可以译为“自动控制装置”。有类似语义变化的词语还包括：defense→defenses(防御→防御体系)，comfort→comforts(舒适→舒适的条件)，luxury→luxuries(奢侈→奢侈品)。

译文：

无线传感器网络对于温室内自动控制装置的管理十分有用。

对于一些半科技词语，则在阅读过程中首先要判断该词语是否处于专业语境，如果不是，则按普通词语处理；如果是，则要判断是什么专业，再选择相关的语义。例如：

- A few bits of information that the FBI requested from the CIA, had it been granted, could well have stopped the “9 · 11” attacks.
- An ELF wave can transmit only a few bits of information per second.

分析：

bit 作为普通词语，可以指“少量”，而在计算机领域，则指“比特”，是二进位制信息单位。这两句话中，前者显然没有专业背景，那么 bits of 就可以理解为“少量的”；而后一句相关的术语 ELF wave 以及 information 这两个同一语义场的词汇已经明显提示了专业语境，因此，在此 bits 只能作为“比特”来理解。

译文：

- 如果联邦调查局果真得到了它向中央情报局索取的那些情报，那么“9 · 11”袭击事件也许会被制止。
- 极高频电波每秒只能传送几个比特的信息。

又例如：

- In copper and most pure metals there is an appreciable increase in resistance with an increase in temperature.
- Ice floats because it is not as dense as water.
- The bottom layers are rich in nickel, a strong and tough material.

分析：

第 1 句中 resistance 作为普通名词，可以指“阻力、抵抗、反抗”。这句话明显是具有专业背景的，它是在分析金属的特性，那么显然应该选择的意思只能是“电阻”。第 2 句明显是在分析对比冰与水的物理特性，那么形容词 dense 可以转性译成名词：“密度”；第 3 句涉及材料学，strong 和 tough 这两个形容词则对应了材料学中的 strength(强度) 与 toughness(韧性)。

译文：

- 铜和大多数金属的电阻随着温度的升高会明显增大。
- 因为冰的密度比水小，所以它可以浮在水面上。
- 底层含有大量的镍，一种强度高、韧性好的材料。

除了考虑应用的专业领域、同一语义场的参照类别、语法线索外，译者还应考虑中文的搭配习惯，以便获得最佳的表达效果，切不可一味生搬硬套字典释义。

例如：make 一词的基本意思是“制造”，但跟不同的词语搭配，就要根据中文的习惯翻译成不同的词语，例如：make steel 就不能译成“制造钢”，而应该是“炼铁”；make a road 应该是“筑路”；make cloth 则是“织布”。又如：

Using drugs and surgery, doctors can only encourage tissues to repair themselves. With molecular machines, there will be more direct repairs.

分析：

这句话 encourage 一词虽然意思为“鼓励”，但“鼓励组织自行修复”这样的表述显然不符合中文习惯，译者应该根据中文的习惯改译为“促进”，则中文就会比较流畅。

译文：

通过药物和外科手术治疗，医生只能促进组织自行修复，但有了分子机器，就能进行更直接的修复工作。

总之，译者在翻译中要做到勤查勤问，不仅查阅词典，还应查阅相关的文献资料及中英文平行文本，请教相关的专业技术人员以确定某个词汇规范、适切的中文表达。因为，不规范的翻译必然会影响到译文的准确性与可接受性。例如：

Tumor cells can be weakened and injected like a vaccine into a mouse. Afterwards, if these same tumor cells, at full strength, are injected into the mouse, the mouse will reject the tumor cells and cancer will not develop. Tumor cells can be weakened and injected like a vaccine into a mouse.

分析：

这句话的专业语境是在讲疫苗的免疫原理，查阅相关资料可知，疫苗有减毒疫苗与灭活疫苗，前者指制剂中的细菌或病毒的毒性减轻，后者指杀死病毒使其完全失去活性。由此可以推断，此处 weakened 不能仅仅按照字面意思直译为“弱化”，而应结合句子的专业语境，采用更加规范化的表达，即“减毒的”；at full strength 则指“未减毒的”。

译文：

肿瘤细胞可以像疫苗一样在减毒后注射进老鼠体内。之后，如果这些同样的肿瘤细胞在未减毒的情况下注射进这只老鼠体内，老鼠就会排斥肿瘤细胞，因而不会患上癌症。

再例如：

Repair machines will be able to repair whole cells by working structure by structure.

分析：

这句话涉及纳米器械在医学康复中的应用，以及细胞修复机如何重塑人体健康。working 在此处不能仅仅直译为“工作”，而是应该译为“作业”，多指工程领域里完成某一项任务。

译文：

细胞修复机通过对分子结构逐一作业来修复整个细胞。

1.4 英汉语区域变体对术语的影响

英语的两大区域变体是英式英语与美式英语，二者在日常用语及科技术语方面都有一定差异，对术语的单一性原则会产生一定影响，例如：美式英语中，汽车术语“变速箱”是 transmission，而英式英语则是 gearbox，如果译者不了解这一差异，则会给翻译造成一定的障碍。

同样，由于地理、政治及历史现实的差异，我国大陆、港澳台地区虽然都使用的是中文，但在表达方式上却存在明显差异，这些差异在术语的表达上也有体现，例如：大陆的“导弹”“航天飞机”“软件”，在台湾则被称为“飞弹”“太空穿梭机”和“软体”；而港澳人士所说的“波箱”“手波”“自动波”则是大陆人指的“变速箱”“手动变速箱”和“自动变速箱”。

因此译者必须了解这些语言变体之间的用词差异，并且根据服务对象的语言背景采取相应的措施，选择准确的译语来应对。

第 2 节　数词、倍数增减及百分数的译法

2.1 数词的翻译方法

数词（numeral）指表示数量或序数的词，包括基数词与序数词。基数词表示数量的多少，序数词表示顺序。在此，主要介绍基数词的翻译方法。

中英文对基数词的表达法有一定差异，体现在：英语中没有万、十万、千万、亿这样的数字单位，所以他们一些大的数字需要借助下一级的单位来表达，例如，万在英文中为 ten thousand；十万则是 a hundred thousand；千万则需要借助百万这个单位来表达，即 ten million，依次类推，亿则是 a hundred million。因此，万以下的、数值较小的基数词可以直译，例如：

The price of this new machine is about a thousand dollars.

这台新机器的价格是 1 千美元。

但对于较大的数字，则需要换算成汉语相应的数字单位。例如：

Only when a rocket attains a speed of 18,000 odd miles per hour, can it put a manmade satellite in orbit.

只有火箭速度达到每小时 1 万 8 千多英里时，才能把人造卫星送入轨道。

The temperature at the sun's center is as high as 10,000,000 ℃.

太阳中心的温度高达摄氏 1 千万度。

不定量数词的译法

不定量数词指表示若干、许多、大量、不少、成千上万等不确定数量概念的词语。多是由 number，score，decade，dozen，ten，hundred，thousand，million 等词后面加复数后缀-s 构成。在翻译不定量数词时，注意准确翻译数值的大小范围。

dozens of 几十，几打
scores of 几十，许多
teens of 十几（13 ~ 19）
tens/decades of 数十，几十
hundreds of 几百，成百上千
thousands of 几千，成千上万
tens of thousands of 数万，好几万
hundreds of thousands of 数十万，几十万
millions of 千千万万，数以百万计
millions upon millions of 数亿，无数
numbers of 许多，若干
例如：

The number of known hydrocarbons runs into **tens of thousands**.

已知的碳氢化合物多达数万种。

数字修饰语的翻译

英语中常用 flat，cool，sharp，exactly，whole，just 等词修饰数字，表示“正好、恰好、整整”等。例如：

He finished the experiment in twenty-four hours **flat**.

他正好用了 24 小时完成这个实验。

英语中常用 about，some，around，round，approximately，closely to，nearly，towards 等词修饰数字，表示“大约、上下、将近、几乎”。例如：

It is **nearly/towards** 4 o'clock.
现在将近4点种。

英语中常用 more than, odd, over, above, exceed, upwards of, or more, higher than, in excess of 等词修饰数字，表示“超过、以上、有余、高于、多于”等。例如：

Pig iron is an alloy of iron and carbon with the carbon content **more than** two percent.
生铁是一种铁与碳的合金，其中的碳含量超过2%。

英语中常用 under, below, less, less than, no more than, short of, off, to, within, as few as 等词修饰数字，表示“少于、不到、以下”等。例如：

The efficiency of the best of these engines is **under/below/less than/no more than** 40%.
这些发动机中，效率最好的也不到40%。

英语中常用 from ... to, between ... and/to 来修饰数字，表示范围，可译为“从……到……”“在……之间”等。例如：

The energy of the fuel wasted by the reciprocating steam engine is **between** 80 **and** 85 percent.
往复式蒸汽机浪费的燃料能量为80% ~85%。

2.2 倍数增加及倍数比较的译法

中英文表示倍数增加的方式有所不同，差异体现在以下方面：

英语中说“增加了多少倍”，都是连基数也包括在内的，是表示增加后的结果；而在汉语里，所谓“增加了多少倍”，则只表示净增的倍数。所以，英语里凡是表示倍数增加的句型，汉译时都可译成“是……的 n 倍”（包括基数在内），或“比……增加（$n-1$）倍”（指净增倍数）。

英文表示倍数增加与比较的常用句式：
increase/rise/grow/go up by *N* times
increase/rise/grow/go up *N* times
increase/rise/grow/go up by a factor of *N*
increase/rise/grow/go up to *N* times
be *N* times as + 形容词/副词 + as...
be *N* times + 比较级 + than
be + 比较级 + than + 名词 + by *N* times
be + 比较级 + by a factor of *N*
以上句式均指：“是……的 N 倍”“N 倍于……”（包括基数在内）
“比……大（$N-1$）倍”（指净增倍数）例如：

The strength of the attraction **increases by four times** if the distance between the original charges is halved.

如果原电荷之间的距离缩短一半，引力就会增加三倍。(或：增大到原来的四倍。)
The substance reacts three times as fast as the other.
该物质的反应速度是那种物质的 3 倍。(或：该物质的反应速度比那种物质快 2 倍。)
Mercury weighs more than water by about 14 times.
水银的重量约为水的 14 倍。(或：水银比水重 13 倍。)

2.3 倍数减少的译法

英语中一切表示倍数减少的句型，英译汉时都要把它换算成分数。英文常说“减少了多少倍”或者“成多少倍地减少”，但汉语只能说“减少了几分之几”或“减少到几分之几”。

英文表示倍数减少的常用句式：

decrease…(by) *N* times
decrease…by a factor of *N*
reduce…to *N* times
N times less than…

中文只能译为：减少了 $N-1/N$ （减至 $1/N$）。例如：

A is 3 times smaller than B.
A 是 B 的 1/3 或 A 比 B 小 2/3。

再比如：

The loss of electricity was **reduced by a factor of four.**
分析：
电损减少了四倍，也就是降至原来的四分之一或减少了四分之三。
译文：
电损减至原来的四分之一。(电损减少了四分之三。)

2.4 百分数的翻译

英文中一些数量的增减也可以下列方式通过百分数来表达：

increase/decrease by *N*% 表示净增减，百分数照译，译为：增加/减少了 *N*%；
decrease to/fall to *N*%：减少到 *N*%；
***N*% + 比较级 + than** 表示净增减，百分数照译。

要注意百分数前的介词，如果是 by 则表示净增减的数量；如果是 to，则表示增至或减至的数量。例如：

Tests have shown that the energy consuming can be **reduced by 30 percent.**
分析：
百分数前是介词 by，那么这句话表示净减数量。因此可译为“降低/减少 30%”。

译文：

试验已经证明，能量消耗可降低30%。

The loss of energy has been **reduced to less than 10 percent.**

分析：

这句话中百分数前是介词 to，则表示减少后的数量，可译为“减少/降低到。”

译文：

能量损耗已降至不足10%。

The steel output is **30 percent higher than** that of last year.

分析：

百分数用于比较级中多表示净增减的数量，因此可译为“增长了30%”。

译文：

钢铁产量比去年增长了30%。

第3节 名词化结构的翻译

科技文体大多是以事实为基础来论述客观事物，要求作者在遣词造句中客观地表达事物特性，避免主观意识，强调存在的事实，而非某一具体行为，而名词、名词性词组正是表达事物的词汇。所以，在普通英语中用动词等表示的内容，科技英语却常用名词来表达，而把原来的施动含义蕴藏在深层的结构里。另外，名词化结构也是实现科技英语无人称性，即避免使用人称代词做主语的重要途径，从而使文章产生客观、简洁明了的表达效果。

3.1 名词化、名词化结构及其分类

所谓名词化（nominalization），从词性转换的角度界定，是“一个在形态上与小句谓语相对应的名词短语”；也有人把它看作是动词或形容词用作名词这一语言现象；还有人从语义功能的角度看，认为名词化指的就是把某个过程或特征看作事物，而词性转换只是这种现象得以实现的一种方式。概括而言，名词化指把动词或形容词，甚至是一个小句转换为名词或名词短语，使之在语法上具有名词的功能并能当作名词使用，但其基本意义仍与对应的动词、形容词或句子表达的意义相同或相对应。名词化具有浓缩信息的功能、增加小句信息量的功能、掩盖或模糊事实的功能，而且其使用与语类有十分紧密的联系。一般说来，语篇的技术性越强，名词化出现的频率越高。反过来说，名词化使用得越多，语篇的技术性就越强。

名词化结构是指名词连用、有修饰成分构成的名词词组以及由动词、形容词、短语或句子等转化成的名词或名词词组。根据修饰词的不同，可划分为单纯名词化结构、复合名词化结构和由动词/形容词派生的中心词构成的名词化结构。

（1）单纯名词化结构（名词迭加型）。

单纯名词化结构指由一个或多个名词修饰一个中心名词构成的名词化结构，属于名词连用的情况，即在中心名词之前用一个或多个名词，它们皆是中心词的前置修饰语。汉译时，

名词迭加型一般采用顺译法。

例如：

water purification system 净水系统
laser noise amplitude modulation 激光噪声调幅
heat treatment process 热处理过程
illumination intensity determination 照明强度测定
computer programming teaching device manual 计算机编程教学仪器指南
depth gage 测量深度的规或尺

（2）复合名词化结构。

复合名词化结构由一个中心名词和形容词、名词、副词、分词及介词短语等多个前置或后置修饰语构成。修饰词相对于中心名词的位置关系是汉英名词性词组在构造上最明显的差别。汉语修饰都是前置式的，而英语的修饰语有些前置，有些后置。多个汉语修饰位次关系一般是：紧靠中心名词的是表示类别、性质的名词、形容词，其次是动词或动词词组，再次是数量词或代词，离中心名词最远的是表示领属、时间、地点的名词和代词。而英语修饰语序可前可后，单个修饰语一般在前，词组、短语、从句在后。英语修饰语序一般来说遵守以下原则：限定词—描绘性形容词—表示大小、新旧的形容词—表示颜色的形容词—表示类别的形容词—表示类别的名词—中心名词。基本语序是：专有性—泛指性—名词；次要意义—重要意义—名词；程度弱—程度强—名词；大—小—名词。意思愈具体、物质性愈强，与中心名词的关系愈密切的就愈靠近中心名词。例如：

acute bacterial peritonitis 急性细菌性腹膜炎
special strengthening filler material 特殊的强化用的填料
low average stress values 较低的平均应力值
a non-uniform temperature field 非均匀的温度场
a mechanically worked surface layer 经过机械加工的表层
a precise differential air pressure meter 一只精密的差动气压表
a reversing variable-displacement reciprocating pump 可反置的变容积往复式工作泵

（3）由动词/形容词派生的中心词构成的名词化结构。

这类名词化结构通常由实义动词派生的名词搭配介词短语构成，在句中充当主语或宾语。一些行为名词与介词后面的宾语有时构成动宾关系，有时可构成逻辑上的主谓关系。例如：

Archimedes first discovered the principle of displacement of water by solid bodies.

分析：

句中的名词化结构 displacement of water by solid bodies 由动词 displace 派生的名词 displacement 加上两个介词短语构成，用来补充说明 the principle，一方面简化了同位语，另一方面 displacement 与 solid bodies 构成逻辑上的主谓关系。

译文：

阿基米德最先发现了固体排水的原理。

Fabrication is by no means a new idea, but in the past, unimaginative uniformity in design, together with limited materials, led to a natural distaste for this form of construction.

分析：

名词化结构 unimaginative uniformity in design 的中心词是 uniformity, in design 是后置修饰语，用来限制中心词，但 unimaginative 并非修饰 uniformity 而是修饰 design; 另一个名词化结构 a natural distaste for this form of construction 暗含一个动宾结构。

译文：

预制并非什么新理念，但在过去，由于设计缺乏想象，千篇一律，加上材料有限，自然使得人们不喜欢这种建筑方式。

3.2 名词化结构的翻译

(1) 直译。

将名词化结构还是译为名词短语。例如：

The development of a new furnace design and the successful introduction into the production line of this high-productivity furnace were the decisive factors.

分析：

这句话中的第一个名词化结构 The development of a new furnace design 表达了强烈的所属，表示“关于……的研发”，可以采用直译的方式译为“新型熔炉的研发”；而第二个名词化结构 the successful introduction into the production line of this high-productivity furnace 则表示将此熔炉引进生产线，中心词 introduction 由动词 introduce 派生而来，暗含了一个动宾结构，可以还原。

译文：

新型熔炉的研发以及将此高产熔炉成功引进生产线是两个决定性因素。

The building of these giant iron and steel works will greatly accelerate the development of the iron and steel industry of our country.

分析：

上句中含有两个名词化结构，分别是 the building of these giant iron and steel works, 由 build 的动名词加上介词短语构成，充当句子的主语，以及 the development of the iron and steel industry of our country, 由 development 加上介词短语构成，这两个名词化结构在此句中均可直译为名词短语。

译文：

这些巨型钢铁厂的建设将极大加快我国钢铁产业的发展。

上文中提到的单纯名词化结构也多用直译的方式处理，例如：

This investigation is concerned with precision grinding techniques.

分析：

这句话中的名词化结构是由一个名词及动名词做修饰语的名词短语 precision

grinding techniques, 直译为“精密磨削技术”。

译文：

这项研究涉及精密磨削技术。

Each of the bits has a different optimum cutting speed.

分析：

bit 用于不同领域，专业意义不同，由 cutting speed 提示，此句话的专业语境应该是机械工程，在此领域该词指“刀头”。a different optimum cutting speed 是一个复合名词化结构，其语序与中文相符，可以直译为“不同的最佳切割速度”。

译文：

每一个刀头都有一个不同的最佳切割速度。

（2）将中心词转译为动词，则此名词化结构转换为主谓结构或动宾结构。

因为许多名词化结构是从实义动词派生而来的名词作为中心词，搭配介词短语构成，其深层意义相当于汉语的小句。在翻译时，可以根据汉语习惯将这个中心词采用转性译法，译为动词，这样就可以将此结构还原成一个主谓结构或者动宾结构来译。例如：

All substances will permit the passage of some electric current, provided the potential difference is high enough.

分析：

名词化结构 the passage of some electric current 的中心词 passage 由动词 pass 派生而来，且暗含了一个主谓结构，即 electric current passes。在翻译时可以将中心词转译为动词“通过”，则本名词化结构可译为“电流通过”。

译文：

只要有足够的电位差，电流便可通过任何物体。

The rotation of the earth on its own axis causes the change from day to night.

分析：

这句话中的名词化结构 The rotation of the earth 暗含主谓结构：the earth rotates, 所以可以将 rotation 转译成动词，将此结构还原。

译文：

地球绕轴自转，引起昼夜的变化。

Television is the transmission and reception of images of moving objects by radio waves.

分析：

本句中名词化结构 the transmission and reception of images of moving objects 的中心词分别源自动词 transmit 和 receive，后接介词短语构成一个名词化结构，暗含了动宾结构，即 transmit and receive images, 按照汉语习惯将名词转译为动词，则此结构就还原为动宾结构，即“发射和接收活动物体的图像”。

译文：

电视通过无线电波发射和接收活动物体的图像。

A careful observation of the specimen shows a beautiful contrast between a deep iron blue of the contracted lung, intersected by grey bands, and the glistening white firm covering of the altered pleura.

分析：

A careful observation of the specimen 暗含了一个动宾结构：observe the specimen carefully，因此，在翻译时可以还原回去，译为“仔细观察样本”。

译文：

仔细观察样本，结果显示，感染的肺部呈现深铁蓝色，间或有些灰色条带，而变形的胸膜上则有一层结实的白色覆盖物在闪烁着，蓝白两色形成了漂亮的反差。

(3) 由于形容词名词化结构大多隐含一个小句，可以通过转性译法将中心词转译为形容词，则此名词化结构就还原成一个汉语的小句。例如：

Fabrication is by no means a new idea，but in the past，unimaginative uniformity in design，together with limited materials，led to a natural distaste for this form of construction.

分析：

本句话中的复合名词化结构 unimaginative uniformity in design 暗含一个小句，即 the design was uniform and unimaginative，可以将 uniformity 转译还原成形容词；第二个名词化结构 a natural distaste for this form of construction，暗含一个动宾结构，将 distaste 转译为动词，natural 转译为副词，就可将此结构转换成动宾结构。

译文：

预制并非什么新理念，但在过去，由于设计缺乏想象，千篇一律，加上材料有限，自然使得人们不喜欢这种建筑方式。

The usefulness of fission as a source of energy depends on man's ability to find or produce fissionable materials and to perfect the means of controlling the process.

分析：

本句中有两个形容词名词化结构，分别是 The usefulness of fission as a source of energy 及 man's ability to，前者在句中做主语，表示一个结果，此结构隐含一个主语从句，即 whether fission as a source of energy is useful or not；后一个名词化结构做动词宾语，表示产生上述结果的条件，此结构中也隐含一个从句，即 whether man is able to find or produce…，将这两个名词化结构还原成句子，只需将两个中心词变性转译为形容词即可。

译文：

裂变是否能成为有用的能源，取决于人类能否找到或生产出可裂变材料，能否完善控制裂变过程的手段。

(4) 将名词化结构译为独立的小句。

这种情况通常出现在名词化结构较长而且较为复杂的情况下。例如：

The slightly porous nature of the surface of the oxide film allows it to be colored with either organic or inorganic dyes.

分析：

The slightly porous nature of the surface of the oxide film 这个名词化结构较长且较复杂，但深层结构隐含着一个表示原因的从句，即 the surface of the oxide film is slightly porous by nature, 与句子后半部分构成因果关系。

译文:

氧化膜表面具有轻微的渗透性，因此可以用有机或无机染料着色。

This position was completely reversed by Haber's development of the utilization of nitrogen from the air.

分析:

本句话中的动词名词化结构 Haber's development of the utilization of nitrogen from the air 相当复杂，它内含了两个动词名词化结构，且第二个名词化结构还包含一个复合名词结构 nitrogen from the air。两个动词名词化结构都暗含着动宾结构，即 develop a method to…及 utilize nitrogen from the air, 可以将两个名词转性译为动词，两个名词结构还原成动宾结构，“发明使用……的方法”和“使用氮气”，整个名词化结构与主句构成因果关系，因此可以将其译为独立的小句，做原因状语。

译文:

由于哈勃发明了利用空气中的氮气的方法，这种局面就完全改观了。

3.3 与名词化结构搭配的衍生动词的翻译

在英语中，由于名词化结构的使用，势必在句法结构中要增加与之搭配使用的动词，这类衍生动词的翻译要酌情处理。

(1) 省译。

有些与名词化搭配的动词意义空泛并不具有实际的动作意义，只起语法作用。这类动词包括: do, have, keep, make, take, pay, show, perform, find, offer 等，这些动词通常可以不予翻译，反而是名词化结构中蕴含的动作意义应予以特别关注，在翻译时要充分表达出来。例如:

This chapter provides a brief review of the theory of modification.

分析:

本句话中的名词化结构 a brief review of the theory of modification 暗含了一个明显的动宾结构，即 review the theory of modification briefly, 可以将中心词转译为动词，还原蕴含的动宾结构，而动词 provide 并无实际含义，因此可以予以省略不译。

译文:

本章简单回顾一下诱发变异理论。

Rockets have found application for the exploration of universe.

分析:

本句话中的动词 have found 并无实际意义，名词化结构 application for the exploration of universe 蕴含着被动结构，即 be applied to explore the universe, 名词化结构中的两个名词都可转性译为动词，谓语动词省略不译。

译文:

火箭已经用来探索宇宙。

The addition of alloying elements is made principally to improve mechanical properties, such as tensile strength, hardness, rigidity and machinability.

分析：

在本句话中，动词 is made 纯粹是与 addition 搭配使用而产生，并无实际意义，可予以省译，而名词化结构 The addition of alloying elements 则应按照其动作意义（to add alloying elements) 来翻译。

译文：

加入合金元素主要是为了改善一些力学性能，诸如拉伸强度、硬度、刚度及可机械加工性，等等。

(2) 直译。

如果与名词化结构搭配的动词为 show，suggest，reveal，indicate，demonstrate，mean 等时，它们大多需要译出，按照字面意思直译就可以。例如：

Oxidation implies the combination of other elements with oxygen to form compounds.

分析：

名词 oxidation 由动词 oxidize 派生而来，表示氧化过程/作用，在句子中做主语，implies 在句子中作谓语，具有实际意义，因此需要直译出来。名词化结构 the combination of other elements with oxygen to form compounds 在句子中做宾语，此结构暗含一个主谓结构，即 other elements combine with oxygen to form compounds, 可以将此结构中的中心词 combination 转译为动词，则此结构就还原为一个主谓结构的小句。

译文：

氧化意味着其他元素与氧元素结合形成了化合物。

However, incomplete community knowledge of the drug's proper use suggests that education efforts may further improve outcomes.

译文：

然而，对该药物的合理使用，社区的人们了解得并不全面，这意味着加强教育可以改善成效。

第 4 节　缩略词的翻译

在科技英语中一些名词、短语、术语及概念会在文中不断提及，尤其是一些专业词汇组成的短语对于读者而言会比较陌生，但由于频繁使用词语全称会增加阅读难度，因此，才会在科技英语中使用缩略词，从而帮助读者快速理解文章的意思。此外，一些专业词语会反复出现，若每一次都使用全称会占据较多空间，并且使文章布局零散、杂乱无序，更增大了读者的阅读难度。使用缩略词可以使文章结构紧凑，从而改善其可读性。因此，了解缩略词的构成及翻译是很有必要的。

4.1 缩略词的构成

把词的音节加以省略或简化而产生的词统称为缩略词，这种构词方法称为缩略法(shortening)。缩略词的构成有以下几种：

(1) 截短词。

英语截短词也称裁减法，是指从某个词的完整形式中删去一个或不止一个音节所得到的缩略词。常见英文截短词有以下几种形成方法：

截除词尾（apocope）

据心理学家测定，一个词语的词首部分给听众的印象最深，所以截短词大多是截除词尾，保留词首来代表原词。例如：

ad←advertisment	广告
auto←automoble	汽车
doc←doctor	医生
dorm←dormity	宿舍
exam←examination	考试
fax←facsimile	传真
gas←gasoline	汽油
limo←limousine	轿车
loco←locomotive	火车头
maths←mathematics	数学
psycho←psychotic	精神病患者
porn←pornography	色情文学
taxi←taxicab	出租汽车
tab←tablet	药片，片剂

截除词首（aphaeresis）

有些词如果截除词尾保留词首，会引起语义不清或与其他词混淆，因此也有截除词首的截短词。例如：

bus←omnibus	公共汽车
cello←violoncello	大提琴
coper←helicopter	直升机
dozer←bulldozer	推土机
drome←aerodrome	飞机场
quake←earthquake	地震
van←caravan	大蓬货车
wig←periwig	假发

截除首尾（front and back clipping）

有个别词截去首尾，保留不在首尾的重读音节。例如：

flu←influenza	流行性感冒
fridge←refridgerator	冰箱
tec←dectective	侦探
dept←department	系、部
scrip←prescription	药方
shrink ← head shrinker	精神病医生
comm ← telecommunication	电信
jack←car-jacking	(对汽车等的) 抢劫 、打劫

截除词腰（syncope）

这样的情况很少见。有些词音节相当多，其中又含有读起来相似的音节，就可能截除一个音节。例如：

fluidics←fluidonics	射流学
symbology←symbolology	象征学

以上四种是单词的截短，词组也可以截短。一种是留下词组中的一个词来表达整个词组的意思，例如：daily ← daily paper（日报）、finals←final examinations（期末考试）；另一种是留下的词再经过截短，例如：pub←public house（酒吧，小旅馆）、zoo ← zoological garden（动物园）。

（2）首字母缩略词。

利用词组中各实意词的第一个字母组合代表该词组的缩短词，就叫作首字母缩略词。从结构来看，首字母缩略词可以用字母代表整个词，例如：

EEC←European Economic Community 欧洲经济共同体

UN←United Nations 联合国

WTO←World Trade Organization 世界贸易组织

FLTA←Foreign Language Teaching Agency 上海外教网

GMT←Greenwich Mean Time 格林尼治标准时间

ASEAN←Association of South East Asian Nations 东南亚国家联盟

也可以用字母代表词的一部分，例如：

TV←television 电视机

CD←compactdisc 激光唱片

有的缩略词还可以和其他词连用，例如：

E-mail←electronic mail 电子邮件

V-Day←Victory Day 第二次世界大战胜利日

从应用来看，首字母缩略词几乎遍及所有领域，应用最为广泛的是科学技术的各个专业，例如：

PVC ← polyvinyl chloride 聚氯乙烯

FRP←fibre glass reinforced plastic 玻璃钢

M←metre 公尺

kg←kilogram 千克

F←fluorine 氟

专有名词也多用缩写，以新闻通讯社为例：

AP←Associated Press 美联社

CCTV←China Central Television 中央电视台

来自拉丁语的首字母缩略词，在英语的书面语中已必不可少：

e. g. ←exampli gratia 例如

i. d. ←idem 同上

i. e. ←id est 即

etc. ←et cefra 等等

viz. ←videlicet 也就是说

v. ←vide 参见

（3）首字母拼音词。

把用首字母组成的缩略词拼读成一个词，就是首字母拼音词。过去运用首字母的缩略词只按字母读音，而现在越来越多的科技术语、组织名称、产品名称等都拼成一个词来读，例如：中文的“家庭办公族”就是根据首字母拼音词（SOHO← Small Office Home Office）音译而来的。日常用语里也出现了首字母拼音词，例如：

丁克族 Dink(←dual income, no kids 无子女的双职工家庭)，尼克族 nilk (← no income, lots of kids 无收入多子女家庭)。

例如机构、组织名称：OPEC [aupek] ←Organization of Petroleum Exporting Countries (石油输出国组织)；UNESCO [ju: jneskau] the United Nations Education Science and Culture Organization 联合国教科文组织。

再如科技术语：laser [ˈleiza] ← lightwave amplification by stimulated emission of radiation (激光)；DOS [dos] ←Disk Operation System 磁盘操作系统。

近年来，首字母拼音词出现了一个有趣的现象，人们有意或无意地把一些首字母拼音词拼写成与现存的词相同的样子，并借用其读音。例如：SALT 不是“盐”，而是“限制战略武器会谈”(Strategic Arms Limitation Talks)，所以译者在翻译时切不可望文生义，一定要根据上下文搞清楚词的所指是什么。

（4）拼缀词。

组成复合词的各词中，一个词失去部分或者各个词都失去部分音节后连接成一个新词，这样的构词方法叫作“拼缀法”。用这种构成方法构成的词叫作“拼缀词”或“合成词”，也叫“混成词”(telescopic word) 或“行囊词”(portmanteau word)。拼缀词实际上是复合词的一种缩略形式，是一种古老的构词方式。在20世纪以前，拼缀词的数量并不大，只是在20世纪以来才特别受到公众的青睐，涌现出大量的拼缀词。虽然拼缀词在英语总词汇量中占的比例不大，但在当代英语中却是一种很有生命力的构词方法。

拼缀词以名词居多。例如：sci-fi 科幻小说←science + ficiton；brunch 早中饭 ←breakfast + lunch；telex 电传← teleprinter + exchange；Eurodollar 欧元←European + dollar。

其他词类的拼缀词为数不多。例如：拼缀动词 guestimate 瞎猜、瞎估计←guess + estimate；拼缀动词 breathalyse 对呼吸的测醉分析← breath + analyse；拼缀形容词 fantabulous 极出色的←fantastic + fabulous。

（5）逆构。

这是一种很有趣的构词方式，虽为首字母缩略语，但排列顺序却相反，例如：RC← conditioned response 条件反应；WG ← gas welding 气焊；SI←International System of Units 国际单位制；CA ← air cooled 气冷式。

逆构的原因很多，有的是避免混淆，有意与另一个顺构的缩略语区别，例如：RAI← International Reading Association 国际读者协会；IRA ← International Recreation Association 国际娱乐协会；有的则来自源语，例如 CE← Communaute Europeenne 法语：欧共体。其英文全称为：European Community；有的用于开玩笑，例如：CMAR = Royal Army Medical Corps 英国皇家陆军医疗部队，为讽刺有些人“不能解决问题，”由 Can't Make Anything Right 或 Can't Manage a Rifle 逆构而成；有的出于保密需要，例如：Freds←Special Detachment of Military Reconnaissance Force 英军空降侦察特遣队，反向缩略为 Freds“弗雷德兄弟”，易记、自然，适于保密。

4.2 缩略词的翻译方法

（1）音译或音译加注法，或音译与意译结合的形式。

科技英语中的很多缩略词对中国读者而言，属于文化空缺词，对这类空缺词大多采用音译法或音译加注的方法来翻译，即根据该缩略词的发音来译，或音译基础上在词尾再加上类属词来翻译。例如：

AIDS(Acquired Immune Deficiency Syndrome 获得性免疫缺陷综合征）艾滋病

Rifle 来福枪

OPEC 欧佩克（Organization of Petroleum Exporting Countries 石油输出国组织）

N (newton 力学单位）牛顿

A(amper 电流单位）安培

Radar (Radio detection and ranging 无线电定位与测距）雷达

laser (Light amplification by stimulated emission of radiation 受激辐射光放大器，曾音译为镭射）

还有一些计量单位的术语多采用国际通用的罗马化缩略语形式，也以音译意译结合的形式来翻译，例如：

kN(kilonewton) 千牛顿

kg(kilogram) 千克

(2) 直译/意译。

在词语翻译的层面，直译与意译基本是一个意思，即按词语的字面意思来翻译。例如：

flu (influenza) 流感

laser (light amplification by stimulated emission of radiation) 受激辐射光放大器，意译为激光

MTBF (mean time between failures) 平均故障间隔时间

QC (Quality Control) 质量控制

MCS(Missile Control System) 导弹控制系统

同样，英语截短词也可根据原词的词义直译，如 ad“广告”、dorm“宿舍”、gym“体育馆”、demo“示威、示范”、memo“备忘录”等。即使裁截一个词素仍按原词翻译，例如：fiche 仍译为“微缩胶片”，miniskirt 的截短词 mini 仍意译为“超短裙”或音译加意译，译为“迷你裙”。同理，gas station 中的 gas 是 gasoline 的截短词，不能译成“气体”，而是“加油站”。

许多通过拼缀法构词的缩略词大多是按照其原词的字面意思直译出来的。例如：

Smog (smoke + fog) 烟雾

sci-fi(science + ficiton) 科幻小说

Hifi (high fidelity) 高保真

Telecast(television + broadcast) 电视直播

Op amp(operational amplifier 运算放大器) 简称“运放”

英语首字母缩略词通常采用意译的方法译出全称，例如：

DI(Data Input) 数据输入

EGF(Epidermal Growth Factor) 表皮生长因子

GUI(Graphical User Interface) 图形用户界面

HIV (Human Immunodeficiency Virus) 人体免疫缺损病毒

HDV (High Definition Video) 高清晰度影像术

组织机构名称缩略词一般也用意译译出全称，例如：

IDA(International Development Association) (联合国) 国际开发协会

UNCTAD(United Nations Conference on Trade and Development) 联合国贸易和发展会议

OECD(Organization for Economic Co-operation and Development) 经济合作和发展组织

有些还可根据汉语缩略规则赋以简称，例如：

ASEAN(Association of Southeast Asian Nations) 全称：东南亚国家联盟；简称：东盟

COMECON(Council for Mutual Economic Assistance) 全称：经济互助委员会；简称：经互会

EEC (European Economic Community) 全称：欧洲经济共同体；简称：欧共体

PLO(Palestine Liberation Organization) 全称：巴勒斯坦解放组织；简称：巴解组织

个别组织机构名称除意译外也用音译，例如 OPEC (Organization of Petroleum Exporting Countries) 石油输出国组织，亦称"欧佩克"。

(3) 零翻译。

零翻译指原本应该翻译，但是由于某些原因而没有翻译，使得源语的词汇直接进入目的语中的现象。例如：SARS（Severe Acute Respiratory Syndrome 非典型性肺炎），我们日常生活中经常可以听到 GPS（Global Positioning System），BBS（Bulletin Board System），MV (Music Video)，DJ（Disk Jockey）等各个领域的英文缩略词在生活中也广泛使用。这是因为随着国民素质教育水平的提高，一些英语的缩略词已经进入我们的日常生活中，而成为我们语言中的词语，与意译出来的译名相比，它们更加言简意赅，更符合缩略词的性质。对于一些较长的缩略语，在第一次出现时，可以照搬英文缩略词同时括号里标注出其英文全称并意译出来，让读者了解其意义。在下文中再次出现时就可以采用零翻译的形式处理。

(4) 零翻译 + 意译或零翻译 + 注释。

有些缩略词原形冗长复杂，意译出来并不简练，有失缩略的意义；然而，采用零翻译的形式照搬进目的语，又会增加读者的阅读困难，因此，可以考虑折中的办法，采用零翻译加意译或者零翻译加注释的方式处理。例如：MOSFET（Metal oxide semiconductor field effect transistor）全称为"金属氧化物半导体场效应晶体管"，采用这种中西合璧的方式简化为"MOS 场效应管"。例如：

BASIC(Beginner's All Symbolic Instruction Code 初学者通识指导性编码)：BASIC 语言

SIM(Subscriber Identification Module 用户身份鉴别模件)：SIM 卡

IP-phone(Internet Protocol Phone 网间协议电话)：IP 电话

U-disk (USB Flash Disk 适用于通用串行总线接口的高容量移动存储设备)：U 盘

P T E(Peculiar Test Equipment)：P T E 装备

第 5 节 词语的替代与省略的翻译

重复是英语行文中的一大忌讳，为了避免词语重复，英语中常常使用相应的替代形式，或者对重复的词语进行省略。一般而言，在句子内部用不同的词语替代重复的内容是有必要的，但在句子间，重复提及的内容如何替代或者省略则是相对自由的。

5.1 英语中的替代及其翻译方法

英汉语之间在替代（substitution）与省略（ellipsis）的应用方面存在较大差异，汉语民族倾向于使用重复手段，而英语民族却常用替代与省略来回避重复。这就意味着两种语言间某些貌似对应的成分不一定能以直译的方式移植到对方语言中去。也就是说，英语中的替代与省略形式不宜简单地照搬到汉语中。当然，中文常用的词语重复也不宜在英文中重现。因此，译者如果不了解两种语言在此方面存在的差异，就不可能翻译出符合目的语表达习惯的译文来，甚至在翻译的理解阶段就对特定的替代或省略形式产生误解。

用替代的形式（pro-forms）来代替句中或上文已经出现的词语或内容（shared words or contents）是英语说话或写作的一项重要原则。根据替代项的语法功能可以划分不同类型的替代：名词性替代、动词性替代、分句性替代。

（1）名词性替代。

在英语中重复提及某个事物或概念时，往往需要采用人称代词（you，he，she，they，me，him，her，them 等）、指示代词（this，that，these，those，its，their 等）、关系代词（that，who，whom，whose，which，where，what 等）、不定代词（it，one，ones 等）与名词配合使用。而汉语如果需要重复提及某个事物或概念时，大多是重复使用实指名称。因此，英译汉时，有时替代形式可以直译出来，尤其是替代形式为人称代词时，例如：

Heat shock proteins (HSPs) are a group of proteins that are present in all cells in all life forms. They are induced when a cell undergoes various types of environmental stresses like heat, cold and oxygen deprivation.

分析：

这个例句中人称代词“They”替代的是上句话中提到的“Heat shock proteins”，直译出来完全可以。

译文：

热应激蛋白是一组蛋白质，它们存在于一切生命形式的所有细胞中。细胞处于不同形式环境压力（如遇热、遇冷或缺氧）下时，就会诱使它们产生。

但是，很多此类替代往往不能译为相应的汉语指代形式，而必须首先搞清楚代词的所指是什么，再按其真实所指译出来。例如：

Larger inductances would cause a steeper line, and smaller ones a shallow.

分析：

这句话中“ones”替代上半句中已出现过的“inductances”，在译文中则需重复该词。

译文：

电感越大，则斜率越大；电感越小，则斜率越小。

The connecting rod transmits the up-and-down motion of the piston to the crankshaft, which changes it into rotary motion.

分析：

原文中的“which”替代“crankshaft”，“it”替代的是“the up-and-down motion”。译文中分别按其实质所指译为“曲轴”及“上下运动”。

译文：

连杆将活塞的上下运动传递到曲轴，曲轴将上下运动转换成旋转运动。

These challenges have led those who believe in the theory to search for more concrete evidence which would prove them correct. From the point at which this book leaves off, many have tried to go further and several discoveries have been made that paint a more complete picture of the creation of the universe.

分析：

本句话中的不定代词“many”替代上句话中提到的“those who believe in the theory”，在此可以还原为“许多人”。

译文：

宇宙大爆炸理论自诞生以来就不断受到人们的质疑，从而促使那些相信该理论的人们去寻找更确切的证据，以证明他们是对的。在本书完成时，许多人又做了更深入的研究，而且获得了多项发现，使人们可以更加全面地了解宇宙的起源。

(2) 动词性替代。

动词性替代指以替代的形式（包括替代词及句型）来取代谓语动词（词组）以回避重复。这类替代形式包括：代动词（do 或 be）、复合代动词（do so，do it 等）及替代句型（and so + do/be + 主语以及 as + do/be + 主语）。此外，代词 this 或者 that 虽然从逻辑上讲只能代替名词或者动名词，但也可以间接代替动词结构。需要注意的是，汉语中重复动词的情况要远远多于英语，因此，英译汉时，英语的动词替代形式往往需要还原为其所替代的实意动词，但并不是简单地重复实意动词，有时需要更换表达方式，或者根据汉语习惯予以合并省译。例如：

Every solid object will reflect a sound, varying according to the size and nature of the object. A shoal of fish will do this.

分析：

上句中的“do this”代替前一句话中的“reflect a sound”。在译文中可以还原实意动词的意义。

译文：

任何固体都会反射声音，反射的声音因物体的大小和性质的不同而不同。鱼群也反射声音。

Cheetahs vary, and so do the sequence of their genomes.

分析：

代动词“do”替代了实意动词“vary”，替代句型“so do the sequence of their genomes”可理解为“the sequence of Cheetahs' genomes also vary”，但“vary”不宜在一句话里重复两遍，否则译文会显得比较啰嗦，可以使用同义词来表述。

译文：

每只猎豹都各不相同，猎豹基因组的序列也因个体而异。

Light waves differ in frequency just as sound waves do.

分析：

在本例句中，代动词“do”替代了上半句中的谓语结构“differ in frequency”，本句话可理解为：“Sound waves differ in frequency, light waves also differ in frequency”，根据汉语的习惯，这两句话的意思可以合并起来，译为“同……一样，主语也……”，这样代动词“do”就可省译。

译文：

同声波一样，光波的频率也不相同。

The manual transmission requires use of a clutch to apply and remove the engine torque to the transmission input shaft. The clutch allows this to happen gradually so that the car can be started from a complete stop.

分析：

“this”的意义显然与前一句话中的“apply and remove the engine torque to the transmission input shaft”相关，可以视为是该动词结构的名词化结构的替代形式。但在汉语中不宜将“this”简单直译，而应该通过增词的方式译为“这个过程”。

译文：

手动变速器必须配合离合器使用，以便向变速器的输入轴施加或取消发动机扭矩。使用离合器可使这个过程逐渐展开，这样汽车就是在完全静止的状态下也可以起动。

（3）分句性替代。

英语中常用 which 来代替前面的分句，或用 this 代替前面的句子。此外，用 it 来替代后续的 that 从句或不定式结构以及用 as + do/be 或者 so + do/be 来替代分句的情况也比较常见。在翻译过程中这些分句的替代形式可根据其具体指代的内容进行灵活处理。例如：

A sound made by tapping on the hull of a ship will be reflected from the sea bottom, and by measuring the time interval between the taps and the receipt of the echoes, the depth of the sea at that point can be calculated. So was born the echo-sounding apparatus, now in general use in ships.

分析：

在这个例句当中，“so”显然替代的是前句话中提到的整个情形，即如何利用回声探测海底深度，在句子中做方式状语。因为上文已经将原理交代得很清楚了，在此，“so”可以直译处理。

译文：

敲打船体发出的声音会从海底反射回来，测出回声间隔的时间，便可计算出该位置海洋的深度。这样就产生了目前各种船舶上普遍应用的回声探测仪。

For decades it has been known that animals can be “vaccinated” against cancer. This is how it works: Tumor cells can be weakened and injected like a vaccine into a mouse.

分析：

前句话中的“it”所指的是后面的 that 从句“that animals can be‘vaccinated’ against cancer”。后一句中的 it 指代的是上一句话中的疫苗接种，this 指代的是其后半句的内容，“Tumor cells can be weakened and injected like a vaccine into a mouse.”，第一句中的替代词“it”可以省译，只要译出其所指的 that 从句即可；第二句中的“it”可以转译为形容词性物主代词，后面的动词转译为名词“工作原理”，“this”则处理为副词。

译文：

几十年来，人们已经了解到动物可以通过“接种疫苗”来抗癌，其原理如下：肿瘤细胞可以先减毒，之后像疫苗一样注射到老鼠体内。

Liquid water changes to vapor, which is called evaporation.

分析：

在这个例句中，“which”替代的是前一个分句，即“Liquid water changes to vapor”，

"which"可以将其所指代的分句译出，做后半句的主语，而"which"可省译。

译文：

液态水变成蒸汽的过程称作蒸发。

As is very natural, a body at rest will not move unless it is acted upon by a force.

分析：

在这个例句中，"as"替代的是后接的句子（"a body at rest will not move unless it is acted upon by a force".）并在句中做主语。翻译时可以译出它所替代的句子，并将其转译为指示代词"这"。

译文：

静止的物体除非受到外力作用才会移动，这是很自然的。

It is possible to improve the antichatter characteristics of a machine tool by incorporating extra damping.

分析：

在本句话中，代词"it"替代的是后面的不定式"to improve the antichatter characteristics of a machine tool"，翻译时可以译出其替代的不定式的内容，"it"可以省译。

译文：

采用外部减震来改善机床的抗震性能是可行的。

We think it important that theory must be combined with practice.

分析：

在本句话中，"it"替代的是后接的 that 从句"that theory must be combined with practice"，翻译时可以将其所替代的从句译出，"it"转译为指示代词"这"。

译文：

我们认为理论必须结合实践，这点很重要。

5.2 科技英语中同义词替代及近义词复现的翻译

英语中指称某一事物的名词往往可以用同义词来替代或者近义词复现。同义词替代就是用不同的名称来指称同一人或者事物。在一些场合，一些约定俗成的词语已经被人们接受为某个事物的同义词。同义词替代在科学家传记、科技新闻报道、科幻小说等涉及科技内容的文章中，使用较多。近义词复现指上义词（又称概括词 generic term）和下义词（又称下属词 specific term）之间的相互替代，广泛应用于各种科技文章中。大多数时候，近义词复现多是下义词被上义词替代，例如，用 vehicle 或者 car 替代 a Peugeot（一辆标致牌汽车），用 the metal 代替某种具体的金属等。在这类替代中，定冠词 the 往往是最明显的指称形式变换的标志。

译者在翻译时首先要搞清楚这些指称形式的所指对象，切不可只是望文生义，机械地按字面意思处理。在翻译时，要根据表达的逻辑性及读者的理解力对这些指称形式灵活处理。例如：

Until such time as mankind has the sense to lower its population to the points where the planet can provide a comfortable support for all, people will have to accept more "unnatural food".

分析：

在本句话中出现的近义词复现是用 the planet 来指代地球，只需译出其真正所指即可，但不宜按字面意思直译。

译文：

除非人类终于意识到要把人口减少到这样的程度：使地球能为所有的人提供足够的饮食，否则人们将不得不接受更多的人造食品。

High carbon steel is easily heat-treated to produce a strong and tough part. The material has a carbon content above 0.80 percent. It finds wide use in hand tools, cutting tools, springs and piano wire.

分析：

本句话中的 The material 与前一句中的 High carbon steel 显然是同指关系。可以直译为"这种材料"，或者重复与其有同指关系的词语，或者用代词"它"来替代。

译文：

高碳钢易于进行热处理，热处理后可制成强度高、韧性好的零件。这种材料（它/高碳钢）的碳含量高于0.8%，可广泛用于制造手动工具、机床加工的切削工具、弹簧和钢琴线。

5.3 科技英语中省略的翻译处理

省略（Ellipsis）是英文中避免重复的常用手段，就这点而言，其实质与替代是一样的。省略也可称为"零替代"，省略的信息为"不言而喻"的信息，即通过上下文可以获取的信息，如再次出现就会导致行文的拖沓。省略的前提是句子意思完整无缺，不能引起争议和歧义。省略可以是承接前文的内容省略（承前省略），也可以是承接后文的内容省略（后指省略）。此外，省略的内容要能够还原，还原后语法上要求结构对应，逻辑关系正确。

省略分为名词性省略、动词性省略和分句性省略。而在科技英语中大多是前两种类型的省略。在翻译前，译者首先要分析是何种类型的省略，省略了哪些内容，在翻译时再将省略的内容根据汉语的习惯增译出来。

（1）名词性省略。

The best conductor has the least resistance and the poorest the greatest.

分析：

省略的原则一般是省略相同的成分，据此分析，这句话中 the poorest 及 the greatest 后面分别省略了 conductor 和 resistance。此外，the poorest 后还省略了动词 has。还原后的句子应该是 The best conductor has the least resistance and the poorest conductor has the greatest resistance。这句话译者显然不能简单直译，因为译出来也不符合中文表达习惯。译者只能将原句还原后再直译。

译文:

好的导体电阻较小，差的导体电阻较大。

This year the production of TV sets in our plants has increased by 20%, and of tape recorders by 15%.

分析:

这句话中后半句省略了名词 the production 及动词 has increased, 还原后的句子为: the production of tape recorders has increased by 15%, 将省略部分在译文中增补出来即可。

译文:

今年我们厂电视机的产量增长了20%，录音机的产量增长了15%。

(2) 动词性省略。

Another major difference between the two designs is the way the trains levitate. The two systems use opposite ends of the magnet to lift off. One is using attractive force, and the other, repulsive force.

分析:

本句中 the other 后省略名词 system 及动词 is using, 还原后的形式为 the other system is using repulsive force.

译文:

两种设计的另一个重大区别是，列车的悬浮方式不同。这两种系统分别利用了相反的磁极使列车悬浮：一个利用的是吸引力，另一个则用的是斥力。

第6节 专有名词的处理方法

专有名词（Proper Noun）是表示人、地方、事物等特有的名词，英文中其第一个字母要大写。在科技英语中常常会涉及大量的专有名词，诸如人名、地名、机构名称，文献、刊物、著作等的名称。专有名词的翻译，与其说是译出来的，不如说是通过资料检索查出来的。专有名词处理得当与否，与译者常规意义上的翻译能力并无必然联系，但与其综合背景知识、专有名词的语际转换意识，及获取检索资料的能力有很大关系。

6.1 人名与地名的英译中处理方法

人名与地名是最典型的专有名词，大多采用音译或文字借用的方法来处理。英语及其他西方语言中的人名与地名必须经过音译才能进入汉语。现今，较为规范的做法是：首先，必须确定名称持有者的国籍或民族身份，应尽量遵从其所属民族语言的发音规律。源自不同语言的名称应按不同的音译表来翻译，例如：英汉、法汉、德汉音译表。例如，英语中的 Charles 在汉语中应译为“查尔斯”，而法语中的同形字 Charles 却应该译为汉语的“夏尔”（如 Charles De Gaule 夏尔·戴高乐）。其次，还应确定该译名是否已经在汉语中沿用成俗，如果是，则应沿用旧译，不宜轻易改动。贸然改动，恐怕会引起读者的困惑。缺乏规范或规范多样化是人名、地名翻译中的一大难题，其直接后果是译名混乱。为避免因译名混乱导致指称目标丧失，译名时，除了尽量做到统一，首次译名时还要附上原文的名称。英译汉时，有些名称本来自汉语，但其汉字形式已无从核实，因此，译文中需要用大致同音的汉字代替，但要注明其为音译。例如，英文中的 Chang San 回译到汉语时就产生身份重新辨识的问题，这个人可能是“张三”，也可能是“章三”或“常珊”。因此，在无法确定其汉字形式的情况下，有必要在译名后加括号注明是音译。

6.2 英汉语之间机构名称的相互转换

英语汉语之间机构名称的转换大多采用直译的方式。例如，英语中的 Royal Society 对应的中译名是英国皇家学会；The Nano Science Foundation 指美国国家科学基金会，National Aeronautics and Space Administration 是美国航空航天管理局。但不少机构的名称是通过跨语言重新命名进入目的语的，例如，The HongKong and Shanghai Banking Corporation 与其汉语名称“汇丰银行”之间无论在语义还是语音方面均无对应，是不折不扣的重新命名。因此，在翻译机构名称时一定要勤上网查阅，确保译名正确，切不可擅自采用直译或音译的方式自行处理，而应该采用广为接受的译名。

第 7 节　练习

一、思考题

1. 科技英语中的通用技术词汇及半科技词汇分别指什么？翻译时如何确定这些词汇的意义？

2. 中英文表达倍数的增减与比较有何不同？翻译时该如何处理？

3. 什么是名词化及名词化结构？名词化结构有哪些翻译方法？

4. 缩略词有哪几种构成方法及翻译方法？

5. 科技英语中的替代与省略各有哪几类？翻译时应该如何处理？

6. 翻译人名、地名时应该遵循什么规范？

二、翻译实践

1. 阅读下列句子，根据句子的上下文确定并翻译加粗的科技词语的意义。

（1）Until a reactor at the Chernobyl nuclear power plant exploded on April 26, 1986, spreading the equivalent of 400 Hiroshima bombs of **fallout** across the entire Northern Hemisphere, scientists knew next to nothing about the effects of radiation on vegetation and wild animals.

（2）A brief recording of human speech might contain only a few words, but we can **extract** vast amounts of signal-processing data from the tone of voice.

（3）Any intelligent machine is, at its core, a software system consisting of **modules**, each one a program that performs a single task.

（4）Radioactive isotopes are **variations** of atoms that were part of the dusty cloud from which our solar system was born.

(5) Steel **parts** are usually covered with grease for fear that they should rust.

(6) The lower the bandgap, the more of the sun's spectrum a cell can absorb to **excite** electrons, but the lower the energy each electron will have.

(7) The somewhat speculative claims about the possibility of using nanorobots in **medicine**, advocates say, would totally change the world of medicine once it is realized.

(8) But between now and then, the SLS (a single big rocket for both crew and cargo) could carry people to Earth's moon and an asteroid and send a **probe** to search for life on Europa, one of Jupiter's moons.

(9) Concrete foundations need not always be reinforced, but where the load to be carried is and the bearing capacity of the soil is poor, **heavy reinforcement** is usually necessary.

(10) **Mechanical properties** are the characteristic responses of a material to applied forces.

(11) These cameras offer **fully automated controls** as well as **manual and semi-automated controls** and other advanced features.

(12) Most **electric machining processes** have come about due to the development of space-age metal alloys that are tough and hard to machine.

(13) Ozone measuring devices shall be capable of measuring the concentration within **a tolerance of plus or minus** 3%.

(14) The Nimitiz-class aircraft carriers are the largest warships ever built. With over 6, 000 personnel (crew and aircrew), the carrier has **a displacement of 102, 000t**, and **a flight deck length of 332. 9m**.

(15) A rocket carriage can be moved on rails from **the assembly shop** onto the launching turntable.

(16) Parts of numbers smaller than 1 are sometimes expressed in terms of fractions, but in scientific usage they are given as **decimals**. This is because it is easier to perform the various **mathematical operations** if decimals are used instead of fractions.

(17) The main operations are: to add, subtract, multiply and divide; to **square**, **cube** or raise to any other **power**; to **take a square**, cube or any other **root** and to find a ratio or proportion between pairs of numbers or a series of numbers.

(18) In terms of distance traveled the **probability** of an accident is only about one-tenth of that of the safest forms of ground transport.

(19) The sun produces nuclear energy from hydrogen gas and, day by day, its **mass** gets less, as **matter** is converted to energy.

(20) Repair machines will be able to repair whole cells by **working** structure by structure.

2. 以下列出了一些英语数字前缀的含义。

uni- 一，单一	hexa- 六
mono-一，单一	sept-/hept- 七
multi-多	oct (a) - 八
poly-多	deca- 十，十倍
semi-半	deci- 十分之一，分

hemi-半
di-二
bi-二
tri-三
quadr（i）- 四
tetra-四
penta-五
quin-五
cent- 百
centi- 百分之一
kilo- 千
milli- 千分之一
mega- 百万，兆
micro- 百万分之一，微
nano-毫微，纳米

根据以上数字前缀，将下列英文术语与其相对应的中文术语搭配起来。

（1） semiconductor（2） hemisphere（3） carbon monoxide（4） binary system
（5） monophonic（6） quadrode（7） the Pentagon（8） decimal system
（9） millimeter（10） megabyte（MB）（11） nanotechnology（12） megahertz
（13） quadrangle（14） octal number system（15） heptathlon（16） decibel
（17） monocrystalline silicon（18） kilometer（19） polyester（20） Triode

A. 一氧化碳　B. 二进制　C. 半导体　D. 单声道的　E. 四极管　F. 纳米技术
G. 毫米　H. 四边形　I. 分贝　J. 聚酯　K. 兆字节　L. 八进制　M. 五角大楼
N. 三极管　O. 兆赫　P. 千米　Q. 单晶硅　R. 七项全能　S. 十进制　T. 半球

3. 翻译下列句子，尤其注意数字及倍数表达的翻译。

（1） The volume of the sun is about 1,300,000 times that of the earth.
（2） Pig iron is an alloy of iron and carbon with the carbon content more than two percent.
（3） Helium in the air is a little under 1%.
（4） The motor ran 450 solid days on end.
（5） Mercury weighs more than water by about 14 times.
（6） The sales of industrial electronic products have multiplied six times.
（7） This year, the production of this kind of machine in our plant is estimated to increase to 3 times compared with 1980.
（8） Automation will help us to raise the output of production by thirty percent.
（9） The company quadrupled output to around 20 million tons.
（10） Half-lives of different radioactive elements vary from as many as 900 million years for one form of uranium, to a small fraction of a second for one form of polonium.
（11） The melting point of this metal is greater than that of copper by approximately 2.5 times.
（12） Sound travels nearly three times faster in copper than in lead.
（13） By using the new process the reject was reduced to 3 percent.
（14） The cost of operation decreased by five times.
（15） The equipment under development will reduce the error probability by a factor of 7.
（16） The gravity of the earth is about six times as great as that of the moon.
（17） Iron is almost three times heavier than aluminum.

(18) The loss of energy has been reduced to less than 10 percent.

(19) By the year 2003 the world's annual oil output is expected to fall to 33%.

(20) This type of furnace uses 4 times more electricity than that type does.

4. 翻译下列句子，尤其注意句子中名词化结构的处理。

(1) A brief description of how transistors are manufactured is given here.

(2) The ladle refining time takes on average between 3 and 4 hours depending on the treatment required.

(3) Curved rails offer resistance to the movement of the train.

(4) The choice of a suitable rubbing surface for frictional measurement is not always easy.

(5) The growing realization of the importance of phosphorus in plant nutrition has led to a good deal of attention being focused on the periods of the life history in which the element is of the greatest value to the plant.

(6) Nanotechnology is the study of manipulating matter on an atomic and molecular scale.

(7) The somewhat speculative claims about the possibility of using nanorobots in medicine, advocates say, would totally change the world of medicine once it is realized.

(8) Cells damaged to the point of inactivity can be repaired because of the ability of molecular machines to build cells from scratch.

(9) Finally, the self-replication of cells proves that molecular systems can assemble every system found in a cell.

(10) Therefore, cell repairs machines will free medicine from reliance on self- repair alone.

(11) Most importantly, astronomers using the Astro-2 observatory were able to confirm one of the requirements for the foundation of the universe through the Big Bang.

(12) Although the term was not coined until 1967 by Princeton physicist John Wheeler, the idea of an object in space so massive and dense that light could not escape it has been around for centuries.

(13) It was many hundreds of years before any further significant observations were made about the phenomenon of static electricity.

(14) A knowledge of statistics is required by every type of scientists for the analysis of data.

(15) Key to one dominant factor in automation, the sustained substitution of machine for human muscle and the more intellectual use of man's brain, is the ever-increasing effectiveness of his exploration of energy in benign, imaginative, non-conformative ways.

(16) Increased Fas expression could be an indicator of tumor aggressiveness and poor prognosis of renal cell carcinoma.

(17) The butterfly throttle is a major waste of power.

(18) In some cases, deserts are the creation of destruction of virgin forest.

(19) Single crystals of high perfection are absolute necessity for the fabrication of integrated circuits.

(20) Since some materials are not damaged easily as others, the possibility exists of developing radiation-resistant parts.

5. 判断下列各组缩略词是如何形成的，并查阅出其对应的中译名。

(1) ASCII; NASA; UFO; HTTP; DRAM; CPU; BDC; TDC; IT; A-G missile; EDM process; CCTV system

(2) auto; flu; exam; fax; maths; gas; ad; dozer; cello; quake; loco

(3) BASIC; OPEC; UNESCO; AIDS; SIM; SARS; DOS; laser; radar; modem

(4) hi-fi; telex; sci-fi; botel; urinalysis; spamouflage; smog; chloral; comsat; medicaid

(5) CA; WG; RAI; CMAR; Freds

6. 翻译下列句子，尤其注意句子中的替代或省略形式，并将其以适当的方式翻译出来。

(1) As of August 21. 2008, over 800 manufacturer-identified nanotech products are publicly available, with **new ones** hitting the market at a pace of 3-4 per week.

(2) The somewhat speculative claims about the possibility of using nanorobots in medicine, advocates say, would totally change the world of medicine once **it** is realized.

(3) Therefore, since nature has demonstrated the basic operations needed to perform molecular-level cell repair, in the future, nanomachine based systems will be built **that** are able to enter cells, sense differences from healthy ones and make modifications to **the structure**.

(4) The healthcare possibilities of these cell repair machines are impressive. Comparative to the size of viruses or bacteria, **their** compact parts would allow **them** to be more complex.

(5) The other scale in general use nowadays is the binary, or two-scale, in **which** numbers are expressed by combinations of only two digits, 0 and 1.

(6) In terms of distance traveled the probability of an accident is only about one-tenth of **that** of the safest forms of ground transport,

(7) There are two general aims in the field of air safety. The first is to reduce the probability of a catastrophic accident to an acceptable minimum. The second is to endeavor to ensure that in the event of an accident there is the maximum possibility of survival of **the occupants**.

(8) In establishing safety standards, **those** on the ground under the flight path are a main consideration.

(9) Incidents in these flight phases account for over 70 per cent of all accidents, but **they** are also the **ones** in which there may be a reasonable chance of occupant survival.

(10) If the component parts are mainly produced by the contractor in a yard adjacent to the site, **it** is known as on-site industrialized building.

(11) **It** has been found that certain bats emit squeaks and by receiving the echoes, they can locate and steer clear of obstacles—or locate flying insects on which they feed.

(12) The term "development" encompasses three types of process. First, new cells are produced by division. In the high plant, **this** occurs most **commonly**, although not **exclusively**, in regions called meristems.

(13) **It** is important to grasp from the outset that these three phases of development are not necessarily separated either in space or in time.

(14) Divisions may occur in cells which are actively enlarging, and in certain circumstances even **in cells** which would ordinarily be considered as mature and fully differentiated.

(15) The term "differentiated" can therefore have a comparative meaning as well as an absolute **one**.

(16) The root systems of terrestrial plants perform two primary functions: **one** is the acquisition of soil-based resources (primarily water and dissolved ions) and **the other** is anchorage.

(17) Other root system functions, such as storage, synthesis of growth regulators, propagation, and dispel, can be seen as **secondary**.

(18) A flip-flop is an example of a 1-bit memory, and a magnetic tape, along with the appropriate transport mechanism and read/write circuitry, represents **the other extreme** of a large memory with an over-billion-bit capacity.

(19) The main memory is composed of semi-conductor devices and operates at much higher speeds than **does** the file memory.

(20) There are two broad classifications within semi-conductor memories, the read-only memory (ROM) and the read-write memory (RWM). **The latter** is also called a RAM to indicate that **this** is a random-access memory.

(21) The first, lowermost leaves of a side branch are termed prophylls; in monocotyledons only one prophyll is present and in dicotyledons, **two**.

(22) A positive charge was indicated by a plus sign (+) and **a negative by** a minus sign (-).

(23) The electron is one of the particles that make up atoms, the basic units of matter of **which** a chemical element is composed.

(24) Thales thought water was the beginning of everything; Anaximenes, **air**; and Heraclitus, **fire**.

(25) The revolutionary blending of low-observable technologies with high-aerodynamic efficiency and large payload gives the B-2 important advantages over existing bombers. Its low-observability provides **it** greater freedom of action at high altitudes, thus increasing **its** range and a better field of view for **the aircraft's** sensors.

(26) Isolated columns or stanchions are normally supported on square concrete foundation base. Where such columns are spaced at close intervals, it is often more practical to provide a continuous concrete strip foundation to carry a complete row, **as is done for load bearing walls**.

(27) A light-emitting diode (LED) starts off the detection process. The light that it produces hits a fluorescent chemical: **one** that absorbs incoming light and reemits it at a longer wavelength.

(28) Even though **it** is not so strong as **the earth's**, the moon's gravity does something to the earth.

(29) While polyacetylene may be persuaded to conduct current as well as many metals **do**, this material is unfortunately no good for practical use.

(30) In order that an electric current may flow in a circuit **it** must be complete and must consist of materials **that** are electrical conductors.

7. 阅读下面一段文字，并指出其中黑体的专有名词的中文名称。

Steven Chu was awarded **the Humboldt Prize** by **the Alexander von Humboldt Foundation** in 1995 and was a co-winner of the Nobel Prize in Physics in 1997 for the "development of methods to cool and trap atoms with laser light," shared with **Claude Cohen-Tannoudji** and **William**

Daniel Phillips. He is a member of **the United States National Academy of Sciences**, **the American Academy of Arts and Sciences**, **the American Philosophical Society** and **the Academia Sinica**, and is a foreign member of **the Chinese Academy of Sciences**. Dr. Chu received an honorary doctorate from Boston University when he was the keynote speaker at the 2007 commencement exercises. Diablo magazine, based east of Berkeley in Walnut Creek, California, honored Dr. Chu with one of its annual Eco Awards in its April 2009 issue, shortly after he was nominated as energy secretary. Harvard university awarded him an honorary doctorate during its 2009 commencement exercise.

三、拓展阅读和翻译练习

阅读以下科技语篇，并翻译划线句子以及黑体段落。

Black Holes

by Dave Mosher

Don't let the name fool you: a black hole is anything but empty space. Rather, it is a great amount of matter packed into a very small area—think of a star ten times more massive than the Sun squeezed into a sphere approximately the diameter of New York City. The result is a gravitational field so strong that nothing, not even light, can escape. In recent years, NASA instruments have painted a new picture of these strange objects that are, to many, the most fascinating objects in space.

Although the term was not coined until 1967 **by Princeton physicist John Wheeler, the idea of an object in space so massive and dense that light could not escape it has been around for centuries. Most famously, black holes were predicted by Einstein's theory of general relativity, which showed that when a massive star dies, it leaves behind a small, dense remnant core. If the core's mass is more than about three times the mass of the Sun, the equations showed, the force of gravity overwhelms all other forces and produces a black hole.**

Using radio telescopes located throughout the Southern Hemisphere scientists have produced the most detailed image of particle jets erupting from a supermassive black hole in a nearby galaxy. Scientists can't directly observe black holes with telescopes that detect x-rays, light, or other forms of electromagnetic radiation. We can, however, infer the presence of black holes and study them by detecting their effect on other matter nearby. If a black hole passes through a cloud of interstellar matter, for example, it will draw matter inward in a process known as accretion. A similar process can occur if a normal star passes close to a black hole. In this case, the black hole can tear the star apart as it pulls it toward itself. As the attracted matter accelerates and heats up, it emits x-rays that radiate into space. Recent discoveries offer some tantalizing evidence that black holes have a dramatic influence on the neighborhoods around them—emitting powerful gamma ray bursts, devouring nearby stars, and spurring the growth of new stars in some areas while stalling it in others.

One Star's End is a Black Hole's Beginning

Most black holes form from the remnants of a large star that dies in a supernova explosion.

(Smaller stars become dense neutron stars, which are not massive enough to trap light.) 1. If the total mass of the star is large enough (about three times the mass of the Sun), it can be proven theoretically that no force can keep the star from collapsing under the influence of gravity. However, as the star collapses, a strange thing occurs. As the surface of the star nears an imaginary surface called the "event horizon," time on the star slows relative to the time kept by observers far away. When the surface reaches the event horizon, time stands still, and the star can collapse no more—it is a frozen collapsing object.

Even bigger black holes can result from stellar collisions. Soon after its launch in December 2004, NASA's Swift telescope observed the powerful, fleeting flashes of light known as gamma ray bursts. Chandra and NASA's Hubble Space Telescope later collected data from the event's "afterglow," and together the observations led astronomers to conclude that the powerful explosions can result when a black hole and a neutron star collide, producing another black hole.

Babies and Giants

Although the basic formation process is understood, one perennial mystery in the science of black holes is that they appear to exist on two radically different size scales. On the one end, there are the countless black holes that are the remnants of massivestars. 2. Peppered throughout the Universe, these "stellar mass" black holes are generally 10 to 24 times as massive as the Sun. Astronomers spot them when another star draws near enough for some of the matter surrounding it to be snared by the black hole's gravity, churning out x-rays in the process. Most stellar black holes, however, lead isolated lives and are impossible to detect. 3. Judging from the number of stars large enough to produce such black holes, however, scientists estimate that there are as many as ten million to a billion such black holes in the Milky Way alone. On the other end of the size spectrum are the giants known as "supermassive" black holes, which are millions, if not billions, of times as massive as the Sun. Astronomers believe that supermassive black holes lie at the center of virtually all large galaxies, even our own Milky Way. Astronomers can detect them by watching for their effects on nearby stars and gas.

4. Historically, astronomers have long believed that no mid-sized black holes exist. However, recent evidence from Chandra, XMM-Newton and Hubble strengthens the case that mid-size black holes do exist. 5. One possible mechanism for the formation of supermassive black holes involves a chain reaction of collisions of stars in compact star clusters that results in the buildup of extremely massive stars, which then collapse to form intermediate-mass black holes. The star clusters then sink to the center of the galaxy, where the intermediate-mass black holes merge to form a supermassive black hole.

Source: http://science.nasa.gov/astrophysics/focus-areas/black-holes/

第 5 章　句子翻译的常用技巧

自本章开始，我们将从句子的角度讨论一些科技英语翻译的方法和技巧。第五章将先概述地介绍翻译句子时应掌握的一些常用的翻译方法，这些翻译手段和方法也正是处理第六章所涉及的常用句式结构时通常使用的翻译方法。因此，认真学习第五章的相关知识点对第六章内容的学习和掌握十分重要。

要把一种语言的话语转化成另一种语言的话语，除了要具备较高的语言文字水平和专业知识外，还应掌握一些翻译方法和手段。本章将介绍进行科技英语句子翻译时常用的一些翻译方法，这些方法主要包括：直译法与意译法、顺译法与倒译法、合译法与分译法、转译法和综合译法等。

第 1 节　直译法与意译法

直译和意译是提及翻译时大家都耳熟能详的两种翻译方法。了解如何正确使用这两种翻译方法对翻译实践有着重要的意义。

1.1　直译法（Literal Translation）

所谓直译就是翻译时基本保持原句的语言形式和内容，不做大的改动，同时要求译文语言流畅易懂、表述清晰明白。直译强调的是“形似”，主张将原句内容按照原句的形式（包括词序、语序、语气、结构、修辞方法等）用译语表达出来。从这个意义上来说，翻译时要想做到完全的直译，要求还是比较苛刻的。不过，在科技英语中，某些表述定义、介绍、说明等句子结构较为简单、逻辑概念比较直白的文字，采取直译的方式进行翻译也是可行的。例如：

Physics studies force, motion, heat, light, sound, electricity, magnetism, radiation, and atomic structure.

物理学研究力、运动、热、光、声、电、磁、辐射和原子结构。

Like charges repel, unlike charges attract.

同性电荷相斥，异性电荷相吸。

The basic job of computer is the processing of information.

计算机的基本工作是处理信息。

1.2 意译法（Free Translation）

由于英汉语言表达方式的差异，我们常常发现翻译时要想按照字字对号入座的方式进行直译是比较困难的。很多时候，我们会对原句的词序、句序、语法结构、词义表达等方面进行调整或变动，将原句所表达的内容以释义性的方式用译语表达出来，这种翻译方法就叫作“意译”。意译强调的是“神似”。例如：

The design is likely to be accepted on the condition that the cost is reasonable.

假如成本不高，这项设计方案通过的可能性很大。

Resistors are available either in fixed values or variable values.

电阻器分为固定电阻器和可变电阻器两种。

The beauty of laser is that they can do machining without ever physically touching the material.

激光的妙处在于它能进行机械加工而不必实际接触所加工的材料。

在翻译过程中，直译和意译各有不同的作用，两种方法既有明显的区别，又相互补充，二者是动态统一的。因此我们不应把二者割裂开来对待，而是要将两种方法有机地结合起来并加以灵活运用，具体情况具体分析，以准确、通顺地表达原文的内容为最终的目的。

第2节 顺译法与倒译法

由于不同的语言具有不同的构句方式和表达习惯，翻译时要特别考虑句子的语序问题。有些时候可以按照原句表达方式进行顺序翻译，而有时也必须根据汉语表达习惯对原句语序进行相应调整。

2.1 顺译法（Keeping the Original Sentence Order）

顺译法是指译文的语序与原文一致或基本一致。顺译从某个角度上看和直译有一定的相关性，即如果采用直译法翻译，则句子的词序和语序不会做大的调整，那么这也属于顺译。例如：

When light strikes an opaque object, the light is either absorbed or reflected.

当光照在不透明的物体上时，光或是被吸收，或是被反射。

但是，使用顺译法只是在语序上采取和原文一致的方式，而在其他方面，如语句结构、词义表达和搭配方式等也可能采取意译的方式。例如：

The first is to reduce the probability of a catastrophic accident to an acceptable minimum.

其一是尽可能将空难事故发生的概率降到最低限度。

For example, within the field of civil engineering itself, there are subdivisions: structural engineering, which deals with permanent structure; hydraulic engineering which is concerned with systems involving the flow and control of water or other fluids; and sanitary or environmental engineering, which involves the study of water supply, purification, and sewer systems.

例如，在土木工程自己的领域里，就有下列这些分科：结构工程——研究永久性结构；水利工程——研究水或其他流体的流动与控制系统；以及环境卫生或环保工程——研究供水系统、水净化系统与排污系统等。

2.2 倒译法（Reversing the Original Sentence Order）

有些时候要想完全按照原文语序进行翻译是不可能的，如果牵强硬译，会造成语义不通、不符合汉语表达习惯等问题。这时候，必须对原文的词序或语序进行调整和重新构造后才能完成翻译，这种翻译方法称为倒译法。以下我们将对倒译法中词序调整和句序调整的情况分别予以讨论。

（1）词序调整。

英汉语言不同的思维模式造成了英汉词序表达上的差异。英语的思维模式是由点及面的外展螺旋式，其表达方式是由小到大，由近及远，由轻及重，由弱到强；而汉语刚好相反，其思维呈内螺旋式，即由大到小，由远及近，由重到轻，由强到弱。因此，在进行英汉翻译时不能死板地拘泥原文，必须按照汉语表达习惯，重新对相关词语进行排序。例如：

An electron is an extremely small corpuscle with negative charge which rounds about the nucleus of an atom.

电子是绕原子核旋转且带负电荷的极其微小的粒子。

The address is 3612 Market Street, Philadelphia, PA, 19124, USA.

地址在美国宾夕法尼亚州费城市场街3612号，邮政编码19124.

此外，英语中后置定语、状语及同位语在翻译时也需要进行词序调整。例如：

Air safety refers to precautions **taken to guarantee safety to passengers and freight in flight**.

航空安全指为保证飞行中乘客和货物安全所采取的预警措施。

Scientists believe that the nucleus of an average comet is only a mile or two **in diameter**.

科学家认为普通彗星的慧核，其直径只不过一到两英里。

Physicist Edward Teller, **the father of the hydrogen bomb**, argues that the nuclear weapons are "uniquely designed for defensive purposes" and that "we need to know what the other side is doing and how to defend against it".

氢弹之父，物理学家爱德华·泰勒争辩说，核武器是"专门为防御目的设计的"，而且"我们需要知道对方在干什么，以及如何防备"。

（2）语序调整。

有些时候由于英汉语言语法结构和表达语序的不同，翻译句子时会采用前后倒置的语序对整句的语序进行调整。这种倒译的情况常常出现在形式主语句、被动句、状语从句等句式的翻译中（有关这部分的内容在第六章中还会有所涉及和讨论）。例如：

It cannot be stressed too strongly that mathematics is the language of modern engineering.

数学是现代工程学的语言，这样说怎么都不过分。

Thus, the bending stress is very easily computed.

因此，极易计算出其弯曲应力。

Cells damaged to the point of inactivity can be repaired because of the ability of molecular machines to build cells from scratch.

由于细胞修复机可以从零开始重建细胞，因此它能对受损严重到没有活动能力的细胞进行修复。

第3节　合译法与分译法

在翻译的过程中，有些时候我们可以按照原文句子的表达和构句方式进行翻译，不做大的改动；但在大多数情况下，我们必须考虑到英汉语言的差异，在翻译时对原句的结构进行改变，甚至是重建，从而使译文符合汉语表达习惯。合译法和分译法就是用于改变句子结构的两种常用的翻译方法。

3.1　合译法（Combination）

有时原文由两个或多个分句构成，但翻译时可以仅用汉语的一个句子来表达，这种方法称为合译法。合译法是翻译定语从句及其他从句时常用的一种方法。例如：

The first tool-makers who chipped arrows and spears from rock were the forerunners of modern mechanical engineers.

那些用石块削制箭头和枪矛的初期工具制作者，是现代机械工程师的先驱。

Each stalk bears what looks like one large yellow flower.

每一根花葶头上都好像顶着一朵巨大的黄花。

The most important of the factors affecting plant growth is that it requires the supply of water.

影响植物生长的各因素中最重要的是水的供应。

3.2　分译法（Division）

原文是两个或多个句子，英译汉时有时我们可以仍保留原结构，采取分译法译成汉语的多个句子组合的结构。例如：

The cornea is a part of the eye that helps focus light to create an image on the retina.

角膜是眼睛的一部分，它帮助在视网膜上聚光并产生物像。

No matter what the shape of a magnet may be, it can attract iron and steel.

不论磁体形状如何，它都能吸引钢铁。

另外，特别要注意的是，大量使用非谓语结构从而构成逻辑关系复杂的句式结构是科技英语的一大语言特点；而汉语则通常用短句组合的方式来表达复杂的概念。因此，我们常常会采用分译的方式，将英语原文中的一个词或短语用汉语的一个句子来表达。例如：

The doctor analyzed the blood sample for anemia.

医生对血样进行了分析，以判断是否是贫血症。

With the same number of protons, all nuclei of a given element may have different number of neutrons.

虽然某元素的原子核都含有相同数目的质子，但它们所含的中子数可以不同。

Being cheap and effective makes 2. 4-D a very popular weed killer.

由于价格便宜且效率高，2. 4-D 成为深受欢迎的除草剂。

第 4 节　句子成分的转译

转译（Conversion）是英汉科技翻译中经常采用的对原句进行调整、变通的一种翻译技巧。在前面第三章中我们已讨论过词的转性译法，其实，不仅词类可以转译，句子成分也可以发生转译。在翻译过程中，一个或多个词的词性发生了变化，这个词或这些词在句中承担的语法功能势必也会随之发生变化，由此就会出现相应的句子成分的转换。比如，原句中的动词翻译后转变为译语中的名词，这叫词类的转性译法。如果从句子的角度，原句中的动词本来作谓语，转译成名词后就可能成为主语或宾语等由名词充当的成分，这便形成了句子成分的转译。另外，有些情况下，即使原句中的某个词的词性不发生变化，但由于英汉两种语言不同的遣词造句和表达习惯，汉译后原句中某词所承担的语法功能仍可能变化。比如，同样是由名词所作的语法成分，原句中的宾语经翻译后可能变为译文中的主语，等等。

总之，无论是何种情况，我们需要把握的是，翻译过程中有时需要打破原文的句式结构，对译文的语法结构进行一番调整，使之符合汉语的表达习惯和规范，而不是一味拘泥于原文的行文方式，牵强为之。这时，就会常常使用句子成分转译这种翻译方法。一般来说，句子成分的转译包括主语的转译、谓语的转译、宾语的转译、表语的转译、定语的转译、状语的转译等。另外，在一个句子中，一种句子成分的转译有时也会继而连锁引起其他成分的转译，这一点需要特别注意。

4.1　主语的转译

在翻译时，原句的主语可以被转译成汉语的谓语、宾语、定语、状语等成分。例如：

The result is that there will be a proliferation of smart, connected devices, from palm-sized and table PCs to Web-enabled phones and Auto PCs.

这一切将使得智能连接设施全面普及，包括掌上电脑、台式电脑、网络电话和车用电脑。(主语转译成谓语)

If a **generator** is not provided, a battery system with automatic charging features should be provided.

如果没有发电机，则应配备有自动充电能力的蓄电池系统。（主语转译成宾语）

Even though bearings are usually lubricated, there is **friction and some wear**.

即使经常对轴承进行润滑，仍存在摩擦和一些磨损。（there be 句型，主语转译成宾语）

An automobile must have a brake with high efficiency.

汽车的刹车必须具有高效性。（宾语转译为主语，主语转译为定语）

This machine is simple in design, yet it is efficient in operation.

这台机器的结构简单，但工作效率很高。（状语转译为主语，主语转译为定语）

The side of each leaf looks a little like the jaw of a lion, and our name dandelion comes from the French name of this plant, which means "lion's tooth."

从侧面来看，每一片叶子都好像狮子的下颚，蒲公英的名字正是从这种植物的法语名称而来，意思就是“狮子的牙齿”。（主语转译为状语）

4.2 谓语的转译

在翻译时，一般情况下，英文中的谓语不需要进行特别处理。但是，有时根据具体表达的需要，也会将原句的谓语转译成汉语的主语、宾语、状语等成分。例如：

Compression formats **are devised** to greatly reduce file size, yet retain acceptable quality.

压缩格式的设计是为了大幅度减少文件所占的空间，同时基本保持文件质量不变。（谓语转译为主语，主语转译为定语）

The tensile strengths **vary** tremendously with wire size.

抗张强度随弹簧规格的不同而发生显著的变化。（谓语转译为宾语）

Microscope **continues to** be a very important tool in science today.

显微镜仍然是今日科学中很重要的工具。（谓语转译为状语）

4.3 宾语的转译

英语中的宾语，在翻译时可以转译为汉语中的主语、谓语、定语等成分。例如：

These leaves have broad **points** and very jagged **edges**.

这些叶子的顶端较宽，边缘则呈锯齿状。（宾语转译为主语，主语转译为定语）

Physical changes do not result in **formation** of new substances, nor do they involve **a change** in composition.

物理变化不生成新物质，也不改变物质的成分。（宾语转译为谓语）

Much later Heinrich Hertz demonstrated **radio waves** in a primitive manner.

隔了很久，赫兹用一种简单的方法证明了无线电波的存在。（宾语转译为定语）

4.4 表语的转译

由于英汉两种语言在表达上存在差异，在英语中以表语形式出现的表述结构在汉语中经常可以灵活处理成其他表述结构，如主语、谓语、宾语、定语等。例如：

Rubber is a better **dielectric**, but a poorer **insulator** than air.
橡胶的介电性比空气好，但绝缘性比空气差。(表语转译为主语)

This program was not **popular with** all of the troops.
并不是所有军队的人都支持这个计划。(表语转译为谓语)

The metal may be **fluid, plastic, elastic, ductile or malleable.**
金属具有流动性、塑性、弹性、延展性或韧性。(表语转译为宾语)

In this air mechanics laboratory, few instruments are **valuable.**
在这个空气动力学实验室里，贵重的仪器不多。(表语转译为定语)

4.5 定语的转译

英语中的定语在翻译过程中可以转译成汉语的主语、谓语、状语等成分。例如：

There are two different kinds **of electricity**, which we call positive electricity and negative electricity.

电有两种：正电和负电。(there be 句型中，主语后面由介词 of 引出的定语通常转译为主语)

There are only two important chemical reactions **involved** in the combustion of any fuel, be it coal, wood, oil or gas.

任何燃料的燃烧过程只包含两种重要的化学反应，不论该燃料是煤、木材、油还是煤气。(定语转译为谓语)

值得注意的是，由于定语是表示修饰的成分，所以定语的转译有时和被修饰语成分的转译相关联。比如，如果英语中某一名词转译成动词而作了谓语，那么对该词起修饰作用的，由形容词或分词充当的定语结构就会相应转译为汉语的状语。例如：

These new computers are in **wide** use.
这些新型计算机被广泛地使用。(宾语转译为谓语，定语转译为状语)

A **superficial** glance at the table shows a strong correlation between the two sets of measurements.

粗略看一下这张表就可以发现两组测量数据间有密切的关系。(主语转译为谓语，定语转译为状语)

4.6 状语的转译

状语是用来修饰动词、形容词和副词的句子成分。英译汉中，语句中的状语成分可以转译成汉语的主语、定语等成分。例如：

The root is long, thick, and tapering and it has many small fibres or rootlets growing **from its sides and from the end**.

蒲公英的根呈锥形，又长又粗，两侧与末端生长出很多细小的纤维或细根。(状语转译为主语)

At this critical temperature, changes begin to take place **in the molecular structure of the metal**.

当达到这种临界温度时，金属的分子结构开始发生变化。(状语转译为主语)

In some countries the population declined, and governments actively encouraged people to have more children.

有些国家的人口不断下降，因此政府积极鼓励人们多生孩子。(状语转译为定语)

第5节　综合译法（Integrated Method）

在翻译句子时，有些时候单纯地采用前面小节里学习到的某一种翻译方法无法解决所有问题。这就需要我们对句子进行综合分析，将直译、意译、顺译、倒译、合译、分译、转译等翻译方法灵活地、变通地加以综合运用。使用综合译法，句子的语序可以进行调整，句子结构可以进行改造，特别是有些时候碰到无法直接翻译的词语或表述结构，可以采取意译、引申等方式加以处理。这样，译文不仅句法通顺自然，而且会更符合汉语表达的思维方式和表达习惯。在翻译实践活动中，相比较而言，综合译法的使用更为普遍，尤其在翻译科技英语中出现的复杂长句时，综合译法的适用性更强。

Rocket- research has confirmed a strange fact which had already been suspected: there is a "high-temperature belt" in the atmosphere, with its center roughly thirty miles above the ground.

火箭研究证实了早先人们就怀疑的这样一个奇异的事实：大气层中有一个“高温带”，其中心在距地面约30英里高的地方。

译文分析：

本句采取顺译、合译、转换、分译的综合译法完成翻译。译文句子的整体语序和原句基本一致，没做大的调整，此为顺译；定语从句 which had already been suspected 采取合译的方式译为汉语的“的”字定语结构；in the atmosphere 为状语，翻译时进行了转译，变为汉语句中的主语；对 with its center roughly thirty miles above the ground 这一独立主格结构进行分译，用汉语的一个单句来呈现。

Part adjustment and repair must be performed on a regular basis if an acceptable product is to be the end product.

要使印刷品的质量达到要求，就必须对零部件进行定期的调试及维修。

译文分析：

本句虽然不复杂，但翻译时也进行了倒译、转译、意译等几种翻译方法的综合运用。对于 if 引导的条件状语从句采取倒译，符合汉语的表达习惯；on a regular basis 译为“定期的”，这是英语中的状语转译为汉语的定语；特别要注意，这句中 acceptable 和 to be the end product 这两处很难找到汉语中很匹配的词语表达，如果直译为“可接受的产品成为最终产品”则十分牵强。因此，这里我们要做创造性的处理，将两者合起来意译为“印刷品的质量达到要求”，使译文通顺易懂。

In practice, the selected interval thickness is usually a compromise between the need for a thin interval to maximize the resolution and a thick interval to minimize the error.

为保证最大分辨率必须选用薄层，为使误差最小却必须选用厚层，实际上通常选择介于两者之间的最佳厚度。

译文分析：

本句中采用的翻译方法主要有分译、转译和倒译。翻译本句最难的在于对于句子结构的重造。要是按照原文的思维方式进行顺序翻译，会造成逻辑结构混乱，表达不清的问题。由于汉语习惯的思维方式逐层递进，归纳总结，因此，我们先分译出有关“thin interval”和“thick interval”这两种情况的具体内容，然后再采用倒译，翻译结论性的话语“the selected interval thickness is usually a compromise”。这种思维和表达方式的转换十分重要，是该句翻译的灵魂所在。另外，“selected”原来是主语部分的定语，后转译成谓语；“interval thickness”原为主语，后转译成宾语；“compromise”原为表语，后转译成定语。

第 6 节　练习

一、思考题

1. 直译法和意译法有没有优劣之分？在翻译中应该如何看待直译和意译？
2. 翻译过程中为什么会出现语序的调整？语序调整主要包括几种情况？
3. 合译法和分译法，哪种方法在英译汉翻译时使用得更多？
4. 英译汉过程中为什么会出现句子成分转译的现象？
5. 何为综合译法？综合译法一般在什么情况下使用？

二、翻译实践练习

1. 翻译下面的句子，注意翻译过程中直译和意译的恰当应用。

(1) The igniter combustion often produces hot condensed particles.

(2) Einstein's relativity theory is the only one which can explain such phenomena.

(3) The designer must have access to stock lists of the materials he employs.

(4) The law of reflection holds good for all surfaces.

(5) A brief discussion of each processor subsystem will resolve additional details.

(6) In recent years, a more stable product, Penicillin V, has attracted commercial attention.

(7) The theory facilitated the interpretation and prediction of properties of organic compounds.

(8) Exhaust gases from boilers and vehicles cause air pollution in cities.

(9) Recent evidence gives a strong support to the existence of solid hydrates in the nucleus.

（10） Although the United States opened the way for the commercial exploitation of space, Western Europe is now taking the lion's share of benefits.

2. 翻译下面的句子，注意翻译过程中某些句子语序的调整。

（1） Quantification involves putting the data or pieces of information resulting from experimentation into exact mathematical terms.

（2） Engineering is often defined as making practical application of theoretical sciences such as physics and mechanics.

（3） Fourth is the coordinating function of the mechanical engineering, including management, consulting, and, in some cases, marketing.

（4） Typically the fatalities worldwide are about one-third of those killed on the roads in Britain alone.

（5） The result is that massive galaxies take shape before smaller ones, as is described.

（6） Either or both winding may be excited by alternating or direct currents.

（7） Such work requires not only a clear understanding of mechanical science and an ability to analyze a complex system into its basic factors, but also the originality to synthesize and invent.

（8） He looked at radio signals from two nearby stars while studying interstellar gas with a radio telescope at Greenbank, West Virginia.

（9） The two phenomena, inertia and gravitation, seem so different from each other that one can't help but wonder why these two different ways of measuring mass always give the same answer.

（10） According to Berube, there maybe a danger that a "nano bubble" will form, or is forming already, from the use of the term by scientists and entrepreneurs to garner funding, regardless of interest in the transformative possibilities of more ambitious and far-sighted work.

3. 翻译下面的句子，注意翻译过程中合译法和分译法的使用。

（1） The craftsmen who discovered metals in the earth and found ways to refine and use them were the ancestors of mining and metallurgical engineers.

（2） Without the skills included in the broad field of engineering, our present-day civilization never could have evolved.

（3） And the skilled technicians who devised irrigation systems and erected the marvelous buildings of the ancient world were the civil engineers of their time.

（4） Many of the early branches of engineering were based not on science but on empirical information that depended on observation and experience rather than on theoretical knowledge.

（5） Leaves are to the plant what lungs are to the animal.

（6） High steam carbon ratios were required to prevent carbon formation in the reformer tubes.

（7） Being small and light makes transistors more advantageous to vacuum tubes.

（8） It is not always possible to distinguish them (nonheritable variations) from heritable variations without performing an experiment.

（9） What makes Java intriguing is that it is also a runtime environment embodied in what is called a virtual machine (VM).

（10） Because of air resistance, there is a limit on how fast an object falls.

4. 翻译下面的句子，注意翻译过程中句子成分的转译。

(1) Attempts have even been made to rename the robot.

(2) Up to now, there have been three main options for encoding a message.

(3) Restricted movement led to an interruption of blood flow.

(4) The application of surfactants aims to simplify work processes.

(5) More metal is removed when the feed is increased.

(6) We need frequencies even higher than those we call very high frequency.

(7) Another type of lens is thinner in the middle than at the edges, and is known as a concave lens.

(8) Yet, despite the publicity about huge garbage patches in the sea, most of the ocean's plastic isn't big.

(9) At the bottom it is round and white like a little bag.

(10) These horns are ready to catch any pollen dust that insects may bring.

(11) On the microscale, though, it's still super durable.

(12) Unlike the easy-to-observe impacts of large plastic trash, the effects of microplastics are as subtle and difficult to trace as the fragments themselves.

(13) Copper has very high conductivity.

(14) The root of the dandelion stores up in its thick part much food for the use of the plant, to help it in producing its flowers and seeds.

(15) The temperature coefficient of the resistance is positive for metals and negative for semiconductors.

5. 翻译下面的句子，注意翻译过程中综合译法的使用。

(1) Each flower turns into a seed or grain, which becomes hard and round as it ripens.

(2) If you look at other plants, you will find many of them with leaves that gather the rain-water in this way and carry it to the roots.

(3) The microorganisms that decompose ripped-up bits of wood and seaweed down into simpler organic compounds can't easily digest plastic.

(4) The air near the surface of the earth is warmer than that which is higher up, so when the water-vapour rises very high above the ground, it is cooled.

(5) Inside this tube are five very small dust-spikes, all growing together, and forming another and smaller tube.

(6) In many ways, plastic is the perfect material—we can make it strong and rigid enough to build spaceships and replace bones, or thin and flexible enough to make shopping bags that weigh as much as a nickel but carry up to eight kilograms.

(7) If you boil salt water in a small saucepan until the water has all passed away as steam, the salt that was dissolved in it will all be left behind as a white crust on the bottom of the saucepan.

(8) Children who are fond of country walks, and who use their eyes well, will be sure to notice in early spring little lumps of jelly floating on some pond by the roadside.

(9) But if you pull one of these flower-heads to pieces, you will find that it is really made up

of a great many small separate flowers, which all grow close together on the cushion-shaped top of the flower-stalk.

(10) Each of us drinks about three pints of fresh water every day, and we use a great deal of water for washing, for cooking, and for other purposes.

三、拓展阅读和翻译练习

阅读以下科技语篇，并翻译划线句子以及黑体段落。

Why Cancer?

1. The various human cancers are diseases in which one of the many cells of which the normal body is composed gets altered in such a way that it inappropriately replicates itself again and again, producing millions of similarly affected self-replicating descendant cells, some of which may spread to distant parts of the body and eventually overwhelm it.

For perhaps a century or more, medicine has considered cancers arising from different organs of the body as being in many respects completely different diseases, and over the past few decades it has become clear that not only their clinical manifestations and prognoses but also their causes may differ enormously. So, it makes as little sense to lump together cancers of the lung, stomach, and intestine when considering the causes of cancer as to lump together cholera, tuberculosis, and syphilis when considering infective diseases. In particular, it is not true that how one lives makes no difference to whether or not one gets cancer but "merely determines where in the body the disease will be found." On the contrary, there is no general reason to expect that prevention of one type of cancer will cause any kind of compensatory increase in the onset rate of any unrelated type of cancer among people of a given age.

2. Although there are dozens of types and hundreds of sub-types of cancer, a few types predominate in each country, but the particular types that predominate in different countries may differ. 3. In both the United States and Britain cancer of the lung predominates, due chiefly to cigarettes, followed by cancers of the breast, large intestine, and (in Britain) stomach, none of which are much affected by tobacco. Together, these four types account for more than half of all cancer deaths. 4. So, one chief aim of cancer research should be to devise practical methods for reducing the incidence of one or more of these, particular four diseases, because even large reductions in minor types of cancer can have only minor effects on the total impact of cancer.

The fact that present-day cancer rates do not exhibit any generalized increase (apart from the effects of smoking) does not, of course, guarantee that all, or nearly all, of the recently introduced new chemicals, pesticides, pollutants, and habits are harmless, for, it may be decades before any cancer-causing effects may have become clearly evident. 5. Yet, it does guarantee that to concentrate on the scrutiny only of new environmental agents, to the exclusion of the investigation of long-established aspects of our way of life, is to ignore the possibility of preventing the continued occurrence of that mass of present day cancers that are due to avoidable factors that must have characterized the lifestyle of much of the developed world at least throughout this century.

Source: A comprehensive course book of English-Chinese translation

第6章　科技英语中常见句式的翻译

本章主要讨论科技英语中常见的一些句式结构的翻译技巧和方法。这些常用句式包括：被动语态句、并列句、it引导的形式主语句、各种从句（名词性从句、定语从句、状语从句）、否定句以及复杂长句。

第1节　被动句的翻译

英语中的被动语态是通过助动词“be”和及物动词的过去分词组合而成的。通过前面的学习我们知道，由于科技英语描述的主体往往是客观的事物、现象或过程，而非进行研究的个人，因此，大量使用被动语态是科技英语的一大特征。

由于语言习惯的差异，汉语在语态体现方式上与英语有很大不同，在汉语表达中，被动语态的使用频率远小于在英语中的使用频率。英语中用被动的地方，汉语却可能用主动来表达；有些时候即使使用被动句，也会更多地使用隐性被动句或语义被动句（没有“被”“为”等标志词）。

科技英语被动语态句在汉译时，译者应明悉不同语境中被动结构的具体语义、语篇功能和修辞功能，再根据汉语表达方式和习惯，加以灵活多变的处理。

1.1　将英文的被动句译为汉语的主动句

由于汉语习惯上多用主动句，因而许多被动句在汉译时，如果不是非常强调被动意义，都可以译为汉语的主动句。根据不同的情况，主要有以下几种译法：

（1）英语原文中主语仍译为主语。

Codes of practice or “airworthiness requirements” have been established as a consequence of many decades of experience.

飞行业务守则或“适航性要求”是在累积了几十年的飞行经验上建立的。

Medicine, politics, and social policy are increasingly expressed in the language of the mathematical and statistical sciences.

医学、政治、社会政策等越来越多地使用数学和统计学的语言来表达。

More specialized extensions, such as probability theory and group theory, are now applied to an increasing range of activities, from economics and the design of experiments to war and politics.

数学上更专业化的分支，像概率论和群论，现在的应用范围也在不断地扩大，已从经济学和实验设计扩展到战争和政治领域。

（2）英语原文中的其他成分转译为主语。

The non-medical use of certain drugs is forbidden **in the United States** because they can be dangerous.

美国禁止某些药品用于非医疗方面，因为它们可能会造成危险。

The decimal, or ten-scale, system is used for scientific purposes **throughout the world**, even in countries whose national systems weights and measurements are based upon other scales.

全世界都把十进制，即十进位计数法，应用在科学上，即使在那些以其他进位制为法定度量衡的国家里，也不例外。

A lightweight ball suspended from a string can be charged **by touching it with a glass rod that has itself been charged by rubbing with a cloth**.

与和丝绸摩擦而带上电的玻璃棒接触会使系在细线上的轻质小球带电。

（3）译文中添加主语。

Steel and its alloy will still be taken as the leading materials in industry for a long time to come.

在今后很长的一段时间里，人们会将钢以及合金钢作为工业的主要原料。

Salt is known to have a very strong corroding effect on metals.

大家都知道，盐对金属有很强的腐蚀作用。

When the voltage and the current do not reach its peak value at the same time, they are said to be out of phase.

当电压和电流不是同时达到峰值时，我们说它们不同相。

这种添加“人们”“大家”“我们”等表示泛指的人称代词在译文中做主语的译法，在翻译以 it 作形式主语的被动句中，也常常使用。关于这部分，我们将在第 2 节中详细讨论。

1.2 将英文的被动句译为汉语的无主语句

无主语是汉语句法中特有的形式。在英汉翻译过程中，可以将英文的被动句转化为汉语的无主语句。例如：

For any machine whose input and output forces are known, its mechanical advantage can be calculated.

对于任何机器来说，如果知其输入力和输出力，就能求出其机械效益。

No case has ever been found in which conservation of energy doesn't hold.

尚未发现能量守恒定律不适用的情况。

Methods are found to take these materials out of the rubbish and use them again.

现在已找到从垃圾中提取这些材料并加以再次利用的方法。

1.3 将英文的被动句仍译为汉语的被动句

（1）译成典型的汉语被动句。

如果英文的被动句被动意味很强烈，在汉译时也可保持原被动结构不变，借助使用汉语中表达被动的词语，如“被”“由”“受”“让”“给”等，译为“强被动句”。例如：

Non-conductors of heat can also be called heat insulators.

非导热体也可被叫作热的绝缘体。

Everything on or near the surface of the earth is attracted by the earth.

地球上或地球附近的一切物体都受地球的吸引。

This undertaking is sponsored by federal government, state governments, universities, public television stations and commercial networks.

这项事业由联邦政府、州政府、大学、公共电视台和商业网络联合主办。

（2）转换成其他形式的被动句。

在翻译被动句时，有时可根据上下文、语气和风格等，使用“加以”“予以”“得到”“为……所”等汉语词汇进行转化。例如：

Trajectory data retrieved from radar and satellites contain many errors that have to be cleaned up.

从雷达和卫星回收到的弹道数据有许多误差，必须加以清除。

Other advantages of our invention will be discussed in the following.

本发明的其他优点将在下文中予以讨论。

These circulation anomalies have long been recognized.

这些环流异常现象早就为人们所认识。

第2节　And连接的并列句的翻译

英文的句子分为简单句、并列句和复合句。简单句，即只含有一个主谓结构并且句子各成分都是由单词或短语构成的句子结构，在前面章节的例句中多有涉及，在本章不作详细讨论。

英语并列句是由and，but，or等简单并列连词或either…or，neither…nor，not only…but also等复合并列连词把两个或两个以上的简单句连接在一起的句子。

“and”是在英语中非常常见的一个连词，它可以用来连接词、短语和句子，其最基本的含义是“和”“以及”“并且”。但是，当翻译and连接的句子时，and除了可以表示并列和递进外，还可以表示转折、因果、对比、目的和条件等关系。这时，如果把“and”都翻译成“和”或“并且”，就会造成句子前后语义和逻辑上的问题，从而产生误译。因此在翻译这类句子时，一定要先分析清楚and连接的句子之间的内在逻辑关联，再选择对应的汉语连接词进行翻译。

（1）and 表示顺接或递进。

当 and 连接的两个句子是对等、并列或递进关系时，可以译为“并且”“以及”“而且”“甚至”等，有时也可以省译。例如：

But one thing is certain: energy never disappears and, equally, it never appears from nowhere.

但是有一点是肯定的：能量永远不会凭空消失，同样，它也不会凭空产生。

In terms of distance traveled the probability of an accident is only about one-tenth of that of the safest forms of ground transport, and in terms of journey hours it is comparable.

就行走距离而言，飞机发生事故的概率仅仅是地面最安全交通工具的十分之一，而且在旅程时间方面，它也具有可比性。

Generally, nanotechnology deals with structures sized between 1 to 100 nanometer in at least one dimension, and involves developing materials or devices possessing at least one dimension within that size.

一般而言，纳米技术研究的对象是至少有一维的、尺寸介于 1 至 100 纳米之间的结构体。此外，纳米技术还涉及研发至少有一维、在以上尺寸范围内的材料或器件。

（2）and 表示转折。

当 and 连接的前后句是转折关系时，可翻译为“但”“然而”等。例如：

There will always be some things that are wrong, and that is nothing to be afraid of.

错误的东西在所难免，但并不可怕。

Incidents in these flight phases account for over 70 per cent of all accidents, and they are also the ones in which there may be a reasonable chance of occupant survival.

在上述飞行阶段发生的事故占所有事故的 70% 以上，但也正是在这些飞行阶段乘客能够得以逃生的可能性最大。

Several disadvantages tend to limit the use of hydraulic controls and they do offer many distinct advantages.

液压控制虽然有许多突出的优点，但也存在一些缺陷，使其应用范围受到限制。

（3）and 表示因果。

当 and 用来连接两个因果关系的句子时，翻译时应先分析清楚 and 连接的后句是“因”还是“果”，然后再将其译为表示原因的“因为”或表示结果的“因此”“所以”“从而”等。

and 表示原因

Laser is widely used for developing many new kinds of weapons, and it penetrates almost everything.

激光广泛用于制造各种新型武器，是因为它的穿透力很强。

Aluminum is used as the engineering material for planes and spaceships and it is both light and tough.

铝用作制造飞机和宇宙飞船的工程材料，因为铝质量轻，韧性好。

No one doubts that these electronic brains are, in a real sense, intelligent, and they possess reason, memory, judgment, even creativity and the ability to learn.

"电脑"很聪明，这一点大家都认同，因为电脑能做理性分析、有记忆功能、能判断，甚至具有创造性及学习的能力。

and 表示结果

Sound is carried by air, and without air there can be no sound.

声音靠空气传播，因此没有空气也就听不到声音。

Some speculate that drug abuse, chronic stress, or intestinal parasitic infections may weaken the immune system and allow AIDS a foothold.

一些研究人员认为，滥用药物、长期精神紧张或肠道寄生虫感染，会削弱免疫系统，从而给艾滋病以立足之地。

In addition, invertebrates are discouragingly numerous for comprehensive survey inventories, and they remained the province of amateur specialists long after vertebrate animals became the objects of organized survey.

此外，对于综合性物种勘查来说，无脊椎动物的数量之大使人们望而生畏，因此在对脊椎动物展开有组织的勘查后很长时间，无脊椎动物仍然是业余研究人员的研究对象。

(4) and 表示对比。

and 还可以连接两个对比关系的句子，翻译时常译为"而"。例如：

Most of living creatures are unable to live far outside the region that suits them best, and man can go almost everywhere.

大多数生物一旦远离最适合自身生存的区域就无法存活，而人类却几乎可以去任何地方。

Researchers plan to create computer copies of the atmosphere on the planet Mars, and others want to study the formation and development of galaxies.

一些研究人员计划制作出火星大气层的计算机仿真图像，而另一些研究人员则打算研究银河系的形成和发展。

A collection of data is called a data set, and a single observation a data point.

一批数据叫数据集，而单个观测结果叫数据点。

(5) and 表示目的或条件。

and 连接两个动词时，若后面的是前面的目的，可以翻译为"来""以便"等。若 and 连接的前句是祈使句，并且和后句是条件关系时，译为"则""那么""就"等。例如：

The solution was to place many filters in the system and hope for the best.

当时的解决办法是在系统中放置许多过滤器，以期获得理想的（去污）效果。

We have to replace some of its parts and make this machine more efficient.

我们必须调换某些零件，以使这台机器效率更高。

Change one or more steps and we will improve the quality of the finished products.

若改变一个或几个步骤，我们就有可能提高成品的质量。

第3节 形式主语句的翻译

英语中有一些句子由“it”作形式主语，而把真正的主语放在谓语之后，这种句型叫作形式主语句。形式主语句是英语中特有的一种句型，而且在科技英语文章中尤为常见。虽然这类句子从句子语法类型上归属于名词性从句中的主语从句一类（下一小节内容），但因其在科技英语中的典型性和重要性，我们单独列出，在本小节中进行详细说明。

形式主语句从其句子语态上可分为两大类型，即由“it”作形式主语的主动句和由“it”作形式主语的被动句，其翻译方法也不尽相同。

3.1 “it”作形式主语的主动句的译法

（1）顺译法。

基本按照原句的语序进行翻译，一般情况“it”无须译出。例如：

It is quite remarkable that so much of this rapid development of electrical devices and the resulting industry took place during the nineteenth century, when the nature of electricity was not completely understood.

更不可思议的是，在电的本质未被完全掌握的情况下，19世纪，电气设备和相关工业却取得了飞速的发展。

It was obvious that the possibility that they would succeed in electrolysis was very little.

很显然，他们能够成功进行电解的可能性微乎其微。

It is not clear in my mind why a generator can produce power.

我仍然不清楚为什么发电机能发电。

（2）倒译法。

先译主语从句，再译前面的主句。为了强调，“it”可以译出。例如：

It seemed inconceivable that the pilot could have survived the crash.

驾驶员在飞机坠毁之后竟然还能活着，这看来是不可思议的事。

It is no longer a necessary consequence that an object thrown into the air must fall back to earth.

抛入空中的物体一定会回到地面，这已不再是必然的结果。

It is also important that coolant be equal distributed on both sides of the wheel.

让冷却液均匀地分布在砂轮两侧，这一点很重要。

3.2 “it”作形式主语的被动句的译法

“it + 被动语态谓语 + 主语从句”这种句式在科技英语文章中出现的频率很高，熟练掌

第 4 章　科技英语中典型词语的翻译

英文科技语篇中常常含有大量的科技词汇、数词、倍数增减的表达、名词化结构、缩略语以及各种专有名词，本章将逐一论述这些词语的常用翻译技巧。

第 1 节　科技词语的翻译方法

1.1　科技词汇的组成

从技术性角度划分，科技英语中的科技词汇大致可以分为三种类型：

（1）纯专业科技词（Pure ST Words）。

纯专业科技词指的是那些仅用于某个学科或某个专业的词汇或术语。不同的专业有不同的专业技术词汇或术语。它们的含义精确且使用范围相对狭窄，如：hydroxide（氢氧化物）、anode（阳极）、annealing（退火）、isotope（同位素）、hypophysis（脑下垂体）、diatom（硅藻）、chloride（氯化物），等等。如果不懂得某一特定领域内的一套专门词汇和术语，就无法阅读和翻译该领域的科技文献。这类词语使用范围狭窄，词义单一，其中有很大一部分源自古希腊和拉丁语。它们的特点是含义精确明晰，概念单一狭窄，译者不可以任意自取他法。

（2）通用技术词（Common ST Words）。

有的英语专业词汇属于通用型词汇，广泛应用于不同的专业，而且在不同的专业中往往有不同的意义。例如：transmission 一词可译成："发射、播送"（无线电工程学）；"传动、变速"（机械学）；"透射"（物理学）；"遗传"（医学）。又如 operation 一词可译成"操作"，通用于许多专业，但也有特定的词义，例如，在数学与计算机科学中则指"运算"，在医学中指"手术"，而在军事方面则指"作战"。即便在同一专业中，同一个词又有多种概念。如 power 一词仅在机械力学这个专业里就有"力、电、电力、电源、动力、功率"等含义。由此可见，这一类词汇的特点是一词多义，用法灵活，应用领域广泛。翻译时不能随意猜测，必须仔细研究上下文，慎重地选择恰当的对应词。与此同时，译者积极扩展、更新自己的知识面也是非常重要的。通用技术词大致可以分为两组：

A 组（特点：科技词义单一），例如：

单词	普通词义	科技词义
interactive	相互作用的	交互式
civil	民事/民用的	土木

握这种句式的翻译方法和技巧对从事科技英语的翻译工作者而言十分必要。翻译时，通常将原被动句转译为主动句。“it”一般不译，整个句子常采用顺译的方式，译成汉语的无主语句，或者添加“人们”“我们”“有人”等表示泛指的人称代词或者名词做主语，译为主动句。

（1）译为无主语句。

It has been found that some variations can be passed on from one generation to another and that others cannot.

已经发现，有些变异可以代代遗传，而有些却不能。

It is expected that scientists will use the computer to examine the world's weather system and to make chemical and biological studies.

据预期，科学家们会使用计算机来研究世界气候系统，同时也可进行化学和生物学研究。

It has been predicted that there will be an earthquake here in a few days.

据悉，几天内这里将会发生地震。

（2）增译表示泛指的人称代词作主语，译为主动句。

有时，翻译这类句子时，要根据上下文，适当添加例如“人们”“我们”“有人”“科学家们”等表示泛指的人称代词在译文中作主语，再顺序翻译谓语和主语从句的部分。例如：

In the 1930s, it was found that the atom could be split, releasing huge quantities of energy.

20世纪30年代，人们就发现原子可以分裂并释放出大量的能量。

It is claimed that this natural organic compound can be synthesized by artificial process.

有人认为，这种天然有机化合物可以通过人工合成。

It is always thought that technology is going to make warfighting and a lot of things easier.

我们总是认为，利用技术能够使作战和其他许多事情变得简单易行。

为了提高科技英语翻译的效率，保证译文的准确性，我们将常用的“it + 被动语态谓语 + 主语从句”的用法及一般翻译规律总结如下：

It is said that...据说……

It is reported that...据报道……

It is estimated that...据估计……

It has been predicted that...据悉……

It is expected that...预期……

It is claimed that...有人宣称……

It has been found that...已经发现……

It has been discovered that...业已发现……

It has long been known that...人们早就发现……

It has been observed that...我们已经注意到……

It has been proved that…业已证明……

It had been generally believed that…人们普遍认为……

It is widely acknowledged that…普遍认为……

It is generally accepted that…一般认为……

It was incorrectly believed that…人们错误地认为……

It was thought that…人们一度认为……

It will be appreciated that…我们可以理解……

It can be concluded that…可以断定……

It has been demonstrated that…业已证明……

It should be noted that…应该注意到……

It should be realized that…应该认识到……

It is suggested that…有人建议……

It is recommended that…人们建议……

需要特别注意的是，若主句中的时态是过去时，必要时需增译“过去”“以前”“曾经”“一度”等汉语中表示过去时态的时间副词，否则可能会引起误译。试比较下面这句话的两种译文：

It was thought that all atoms of the same element were exactly alike.

译文 1：人们认为同一种元素的所有原子都是相同的。

译文 2：曾经有人认为同一种元素的所有原子都是相同的。

译文分析：

很明显，译文 2 是正确的。因为如果按照译文 1 用一般现在时态来翻译，那么“同一种元素的所有原子都是相同的”这一观点就会成为客观事实，这与“元素有同位素”的概念不符，是错误的翻译。

第 4 节　名词性从句的翻译

大量使用复合句是科技英语的特点之一。复合句即我们通常所说的英文中的各种主从句，从句对主句起到补充和说明的作用，不能独立存在，只充当主句的一个成分。从理论上来说，英文中除谓语外，其他每一个句子成分都可以用句子的形式构成，从而形成名词性从句、定语从句和状语从句这三大类型的从句。我们将在第 4 节至第 6 节中分别对这三种从句的翻译进行讨论。

英语简单句中的名词可以做主语、宾语、表语和同位语，当将这些名词扩展为句子，就形成了各种类型的名词性从句，即主语从句、宾语从句、表语从句和同位语从句，其功能仍然相当于名词或名词性短语。一般来说，名词性从句主要由关系代词（what，whatever，which，whichever，who，whoever，whom，whose）、关系连词（that，whether，if）和关系副词（when，where，why，how）来引导。

4.1 主语从句的翻译

一般来说，英语的主语从句有两种形式：一种是“主语从句+谓语+其他成分”的形式，主语从句在主句之前；一种是由“it”作形式主语，真正的主语放在谓语之后的形式主语从句，关于这部分内容我们已在上一小节进行了讨论。

对于“主语从句+谓语+其他成分”这种结构的主语从句，汉译时通常采用顺译法，即把主语从句放在句首翻译。例如：

What a motor does is to change electrical energy into mechanical energy.

马达的作用就是将电能转换成机械能。

What makes Java intriguing is that it is also a runtime environment embodied in what is called a virtual machine (VM).

Java(一种新型计算机语言) 迷人之处就是将其运行时环境安置在称作虚拟机的装置上。

Whether that UFO was a spaceship from outer space or just a flock of flying birds still remains a puzzle.

那个不明飞行物是来自外层空间的宇宙飞船或只是一群飞鸟，还是个谜。

That like charges repel but opposite charges attract is one of the fundamental law of electricity.

同性电荷相斥，异性电荷相吸是电学的一条基本规律。

4.2 宾语从句的翻译

英语的宾语从句分为动词宾语从句和介词宾语从句两种，以下分别加以讨论。

(1) 动词宾语从句。

由于英语的动词宾语从句的表达方式同汉语的动宾句式结构类似，因此翻译这类句子并不困难，只需根据上下文或语境的需要，适当调整主句和宾语从句的顺序即可。

顺译法

People used to think that energy and matter were two completely different things. We now know that energy and matter are interchangeable.

过去，人们认为能量和物质是两种完全不同的东西。现在我们知道，能量和物质是可以相互转换的。

Smelting experiments in this furnace demonstrated that it was possible to carry out complete desulphurization of sulphide concentrates autogenously.

该冶炼炉进行的冶炼实验表明，可以实现硫化精矿的自动完全脱硫。

But as these crops begin infiltrating our food supply, environmental and consumer groups have begun to question whether potential risks to the environment and human health have been adequately studied.

但是，一旦这种农作物侵入我们的食品供应中，环境组织和消费者组织开始质疑：对于这种农作物对环境和人类健康潜在的危害，我们是否做了充分的研究。

倒译法

But at that time no one could explain why the path of a planet must be an ellipse.

但是，为什么行星的轨道一定是椭圆的，当时谁也说不清楚。

Scientists are also questioning whether foods with a gene inserted to improve one area of their performance could prove detrimental in another.

是否植入食物中的基因在改善它们某一方面的性质同时，也可能会使食物变得有害，对此科学家们也开始产生怀疑。

另外，动宾结构宾语从句中还存在用“it”作形式宾语的情况，翻译时“it”一般无须译出。

Scientists have proved it to be true that the heat from coal and oil comes originally from the sun.

科学家已证实，从煤和石油中得到的热能都来源于太阳。

One should take it into consideration that the term Ohm cannot be abbreviated as the letter O.

应当注意不能将欧姆一词缩写为字母O.

（2）介词宾语从句。

英语中有些宾语从句是跟在介词之后的，由于这类句式在汉语中没有十分对等的句式，故翻译时应灵活处理。

顺译法

The documentation should explain everything possible about the program **including** how the program runs and what type of data is needed.

文档应详细说明程序可能涉及的一切内容，包括程序如何运行、需要什么类型的数据等。

The heating produced does not depend **on** which way current is flowing.

所产生的加热作用与电流向哪个方向流动无关。

A whale differs from a shark **in that** the former is a mammal whereas the latter is a fish.

鲸鱼和鲨鱼的不同之处在于鲸鱼是哺乳动物，而鲨鱼是鱼类的一种。

注意：

本句中 in that 构成了介词短语，表示原因，意为“因为，在于”。

倒译法

These concerns have led to a debate among advocacy groups and governments **on** whether special regulation of nanotechnology is warranted.

由此，一场是否需要为纳米技术制定特别规范的辩论已经在各倡议团体和多国政府之间展开。

Germs can't be seen **except when** they are observed through a microscope.

除非用显微镜进行观察，否则就看不见细菌。

注意：

本句中 except when 构成了介词短语，意为“除非”。

There seems to be no end to what petroleum can do for man.

看来，石油可造福于人类的用途将是无穷无尽的。

4.3 表语从句的翻译

表语从句，顾名思义就是指在句子中充当表语成分的句子。表语从句相对而言较其他从句更好识别，它位于系动词之后，对主语起到解释的作用。英语的表语从句和汉语的判断句相当，较多情况采用顺译方式，翻译成汉语的“是”字句，但有时亦可根据上下文灵活处理。

（1）顺译法。

One of the important properties of copper is that it conducts electricity better than other materials.

铜的重要特性之一是其导电性能较其他材料好。

The present target for civil transport aircraft is that the overall probability of a catastrophic accident should be no more than about one in ten million flying hours, which is not quite being achieved at present.

民用航空运输现在的目标是灾难性事故发生的总体概率不超过一千万飞行小时分之一，但如今这个目标还没有完全达到。

The amount of solar power falling on one square meter in full sunlight is what is needed to light one quite strong electric lamp.

在阳光充足的情况下，照射到一平方米面积上的太阳能相当于一盏大功率电灯所需的能量。

（2）倒译法。

有时也可先翻译表语从句，再译前面的主语。

One of the important properties of plastic is that it does not rust at all.

塑料不会生锈是其重要特性之一。

Part of the immune reaction against AIDS or any virus is that lymph nodes enlarge.

淋巴结肿大是（人体对）艾滋病或其他任何一种病毒的免疫反应。

4.4 同位语从句的翻译

同位语是用来对前面名词的内容进行进一步解释和说明的句子成分。当同位语这一句子成分由句子来充当，就形成同位语从句。同位语从句常用连词“that”引导，但有时也可以由“what, which, who”以及“when, where, why, how”或“whether, if”等引导。位于同位语从句之前的名词通常是抽象性名词，常用的主要有 fact，thought，theory，idea，hope，news，doubt，evidence，belief，discovery，information 等。

同位语从句与定语从句容易混淆，判断是哪种从句时，首先应明悉，定语从句中的关联词如 that，what，when 等关系代词或关系副词，替代前面的先行词在从句中充当一定的成分；而同位语从句中的 that 或其他关联词是连词，在句中不作任何成分，且没有任何实质性含义，只是起到连接作用。另外，定语从句对先行词起修饰和限定作用，而同位语从句是对先行词的具体内容进行进一步说明和解释的。

同位语从句在翻译时可采用顺译、倒译和转译等翻译方法。

（1）顺译法。

如果同位语从句和先行词表现出来的意义是同等的，那么就可以把同位语从句直接顺译成汉语的同位语形式，先行词和同位语之间添加“就是”“即”或直接用“：（冒号）”或“——（破折号）”连接。例如：

We come to the conclusion that recognizable differences exist today between marine and non-marine forms.

我们所得出的结论是，在今天，海洋生物和非海洋生物之间存在着明显的差别。

There was the possibility that a small electrical spark might accidentally bypass the most carefully planned circuit.

总有这种可能——一个小小的火花，可能会意外绕过最精心设计的电路。

But his findings gave some support to the idea that fusion may be possible without extreme heat.

但他的发现也支持了这样一种观点：在没有极端高温的情况下，核聚变也是可以产生的。

（2）倒译法。

There is no doubt that protons repel protons.

质子相互排斥是无可置疑的。

There can be no question that industrialization does raise living standards.

工业化确实能提高生活水平，这一点是毫无疑问的。

There is absolutely no question that acupuncture is promoted as a treatment for pain.

针灸越来越多地被用来治疗疼痛，这一点毫无疑问。

（3）转译法。

有些时候同位语从句还可以转译成汉语的宾语形式或定语形式。例如：

转译成宾语

Almost everyone who has received education has the idea how fast sound travels.

受过教育的人几乎都知道声速有多快。

There is concern, too, that in our ignorance we may be destroying species vital to fabric of ecosystems on which we depend for our survival.

人们还在担心，由于我们的无知，我们也许正在毁灭那些对于我们赖以生存的生态结构起着关键作用的物种。

Last year, for example, biologists for the first time found evidence that planting genetically modified corn in open fields may kill butterflies who feed on the corn's pollen.

比如说去年，生物学家首次发现证据显示，在开放环境下播种转基因玉米可致蝴蝶死亡，因为这些蝴蝶吸食了这种玉米的花粉。

转译成定语

The problem whether atomic energy can be used on a spaceship has not been solved.

原子能是否可用于宇宙飞船的问题尚未得到解决。

The proposal that more equipment should be imported from abroad is to be discussed at the meeting.

从国外进口更多设备的建议将在会议上讨论。

The theory that diseases are caused by bacteria was advanced by Pasteur, a French chemist.

细菌致病的理论是法国化学家巴斯德提出来的。

(4) "the fact that..." 类型同位语从句的译法。

抽象名词"fact"后接"that"引导的同位语从句中，fact 一词的实质意义并不强烈，有时仅仅起到语法上的关联作用，因此在翻译时通常可以省略不译。例如：

We all know the fact that all elements are made up of atoms.

我们都知道，元素是由原子构成的。

In consequence of the fact that molecules have perfect elasticity, they undergo no loss of energy after a collision.

由于分子具有完全弹性，因此分子碰撞后并没有能量损失。

The name "dark energy" refers to the fact that some kind of "stuff" must fill the vast reaches of mostly empty space in the universe in order to be able to make space accelerate in the expansion.

"暗能量"是指某种"物质"会充溢宇宙大部分空荡荡的空间，使宇宙膨胀加速。

第 5 节　定语从句的翻译

定语从句是各类英语从句中最复杂，并且在英语中，特别是科技英文中使用频率极高的一类从句。它由关系代词（which, that, who, whom, whose, as）或关系副词（when, where, why, how）来引导，在句中作定语，对前面的先行词起修饰或限定作用。定语从句按照主句和从句之间逻辑关系的紧密程度分为限制性定语从句和非限制性定语从句两大类。严格来说，汉语中不存在可以完全和英语定语从句对等的句式结构，因此翻译定语从句存在一定难度。翻译时，除了可以将定语从句译成汉语的前置定语外，更多情况需要译成汉语的非定语成分。

定语从句的翻译方法主要包括：前置合译法、后置分译法和状语从句转译法等。虽然限制性定语从句和非限制性定语从句在结构上不同，但翻译方法的类型却大致相同。只是，由于大多数非限制性定语从句和主句的逻辑关系往往比较疏远，在意义上具有独立性，因此使用后置分译法的情况较多，反之限制性定语从句则使用前置合译法较多。

5.1 前置合译法

若定语从句结构比较简单，且意义与主句关系紧密，翻译时可以将从句前置，翻译成汉语的“的”字定语结构，从而将英语的主句和定语从句合译成汉语的单句。

(1) 前置合译法在限制性定语从句中的使用。

一般来说，大部分限制性定语从句是整个句子不可缺少的一部分，对先行词起到限制和修饰作用。如果定语从句翻译成汉语后不长，就可以前置译成汉语的“的”字定语结构。例如：

The genes might cause allergic reactions in people who never had a reaction to that food before.

这些基因会导致那些以前食用该食物从来不过敏的人产生过敏反应。

All the plants and animals which we know of have to breathe and so can live only on planets which have suitable atmosphere.

我们所知的一切动植物都必须呼吸，所以只能生存在有适宜大气层的星球上。

People who live in the area where earthquakes are a common occurrence should build structures that are resistant to ground movement.

居住在地震频发区的人们应当建造能够抗震的房屋。

(2) 前置合译法在非限制性定语从句中的使用。

有些定语从句虽然在形式上是非限制性的，但其内容却对先行词起到限制作用，而不仅仅是补充作用，这时这类非限制性定语从句也可以前置译成汉语的“的”字定语结构。例如：

The electrons, with their negative charges, revolve about the nucleus, which is always positively charged.

带负电荷的电子围绕带正电荷的原子核旋转。

Perhaps this is the “death ray”, which we often read about in science fiction.

也许这就是我们常在科幻小说中读到的那种“死光”。

The result, which have just been reported to Britain’s Medical Research Council, show that some of the divers who were apparently healthy had areas of brain damage similar to those found in stroke victims.

最近提交给英国医学研究委员会的检查结果报告表明：这些潜水员中有些虽然表面上看起来很健康，但脑部有些区域出现同中风病人类似的病状。

5.2 后置分译法

如果定语从句比较长且结构复杂，将定语从句和主句合二为一会显得冗长拖沓，不符合汉语表达习惯，这时翻译时往往将定语从句还按照原来的语序，放在先行词之后，与主句分开译成分句。对于先行词，可采用重复先行词，或用“他（它）”“这”“该”等词对先行词进行替代的方法进行翻译。

（1）后置分译法在限制性定语从句中的使用。

有些限制性定语从句和主句间虽然没有“，”号隔开，但实质上对先行词并没有绝对的限制性，单独译出也并不会破坏主句的完整性。另外还有些限制性定语从句较长，若按照前置合译的方式处理会造成修饰语过长、层次不清的问题，这时也应采用后置分译的方式进行翻译。例如：

Numbers consist of whole numbers (integers) which are formed by the digits 0, 1, 2, 3, 4, 5, 6, 7, 8, and 9 and by combinations of them.

数由整数构成，整数则由0、1、2、3、4、5、6、7、8、9这些数字及其任意组合构成。

The most ambitious project so far is a transatlantic fiber-optic cable to be built by 1988 that could significantly cut the cost of communication between the United States and Europe.

迄今为止，最为雄心勃勃的工程是将于1988年铺成的横跨大西洋的一条光纤电缆，这条线路将大大降低美国和欧洲之间的通信费用。

Plastics is made from water which is natural resource inexhaustible and available everywhere, coal which can be mined through automatic and mechanical processes at less cost and lime which can be obtained from the calcinations of limestone widely present in nature.

塑料是由水、煤和石灰制成的。水是取之不尽的、随处可以获得的资源；煤是用自动化和机械化的方法开采的，成本较低；石灰是通过煅烧自然界中广泛存在的石灰石而得来的。

（2）后置分译法在非限制性定语从句中的使用。

大部分非限制性定语从句和主句之间关系不十分密切，对先行词只是起到解释和说明作用，翻译这种定语从句多采用后置分译法。例如：

All matter is made of atoms, which are too small to be seen even through the most powerful microscope.

一切物质皆由原子构成，原子本身太小，即使用最大倍数的显微镜也无法看到。

Nanotechnology has been described as a key manufacturing technology of the 21st century, which will be able to manufacture almost any chemically stable structure at low cost.

纳米技术被认为是21世纪一门重要的制造技术，它能够以较低的成本制造出几乎所有化学成分稳定的结构。

When oil wells are drilled, the first material obtained is frequently natural gas, which burns with a hot flame.

钻井时，首先获得的物质常常是天然气，它燃烧时发出炙热的火焰。

请注意，有时 which 引导的非限制性定语从句修饰的不是某个词，而是指代前面的整句，在翻译这种句子时，定语从句的主语通常用“这”来代替。例如：

Like charges repel and opposite charges attract each other, which is one of the fundamental laws of electricity.

同性电荷相斥，异性电荷相吸，这是电学中的一个基本定律。

Moist atmosphere makes iron rust rapidly, which leads us to think that water is the influence causing the corrosion.

潮湿的空气会使铁很快生锈，这会使我们认为水是引起腐蚀的原因。

5.3 转译为状语从句

英语中有些定语从句（包括限制性和非限制性定语从句）从语法的角度上看和主句是定语关系，但从和主句的语义逻辑关系上看，其功能更接近状语，往往含有原因、结果、条件、让步、目的、时间等意义。翻译这类定语从句应当先仔细分析主句和从句之间内在的逻辑关联，然后添加适当的关联词，将具有状语性质的定语从句转译为不同类型的状语从句。

（1）转译成原因状语从句。

The instruments that are light in weight and small in volume are used in this space shuttle.

由于这批仪器重量轻、体积小，故在这架航天飞机上使用。

Transformers cannot operate by direct current, which would burn out the wires in the transformers.

变压器不能使用直流电，因为直流电会烧坏其中的导线。

It is necessary for the manufacturing community to look in detail at the formalized procedures of Group Technology, which can be systematically implemented to obtain greater benefits.

制造界必须详细研究成组技术的各种业已成形的方法，因为系统地采用这些方法可以大大提高收益。

（2）转译成结果状语从句。

Scientists say this could lead to design changes in airplanes that would save hundreds of millions of dollars in fuel costs.

科学家们说，这项研究成果将改变飞机的设计，从而可以节约数亿美元的燃料费用。

The stimulation of nerve receptors causes the blood vessels to dilate, which also facilitates blood flow.

刺激神经受体使得血管扩张，从而促进血液流动。

All matters have certain features or properties that enable us to recognize it from chemical and physical tests.

所有物质都有某些特征或特性，从而使我们可用化学和物理实验来进行鉴别。

(3) 转译成条件状语从句。

A body whose position changes with time is said to be moving.

若物体的位置随时间而变化，则认为该物体在运动。

Alloys that contain a magnetic substance generally also have magnetic properties.

如果合金含有磁性物质，那么该合金一般就具有磁性。

An electric current begins to flow through a coil, which is connected across a charged condenser.

如果线圈和充电的电容器相连，电流就开始通过线圈。

(4) 转译成让步状语从句。

The heart, which weighs only 300 grams, performs a vast amount of work during a person's life.

心脏虽然只有300克，但它在人的一生中却承担着大量的工作。

Potential energy that is not so obvious as kinetic energy exists in many things.

势能虽不如动能那么明显，但它却存在于许多事物当中。

Electronic computers, which have many advantages, cannot carry creative work and replace man.

电子计算机虽然有许多优点，但其工作没有创造性，因此不能替代人的作用。

(5) 转译成目的状语从句。

Petroleum must be moved to a refinery where these compounds can be separated.

必须把石油输送到炼油厂，以便分解这些化合物。

The bulb is sometimes filled with an inert gas which permits operation at a higher temperature.

灯泡有时充入惰性气体，以便能在更高的温度下正常工作。

The military workshops should be supplied with dehumidifiers, where materiel waiting for repairs or ready objects can be connected.

军队修配所应装上除湿机，这样待修军品和修好的军品就可以和除湿机相连进行除湿。

(6) 转译成时间状语从句。

The air that is needed by the fuel for combustion is blown by air jets or fans.

燃料燃烧时，可用气嘴或用鼓风机吹入空气。

Electrical energy that is supplied to a lamp can be turned into light energy.

给电灯供电时，电能就转化为光能。

The quantity of fluid which passes through a given section of the pipe can be measured.

当液体流经管道某截面时，可以测得其流量。

5.4 特殊类型定语从句的翻译

(1) 介词 + which 的定语从句的翻译。

介词 + which 的定语从句可以是限制性定语从句，也可以是非限制性定语从句，翻译时要根据具体情况灵活处理。例如：

The carbon, of which coal largely composed, has combined with oxygen from the air and formed an invisible gas called carbon dioxide.

煤的主要成分——碳，同空气中的氧结合，生成一种不可见气体，叫作二氧化碳。(译为主语)

Scientists are making great efforts to study the ways by which transformation of energy is carried out in plants and animals, which will help the science of bionics.

科学家们正在对动植物体内进行能量转换的方式进行大力研究，这将有助于仿生学的发展。(译为前置定语)

Any change in which no new substance is formed is a physical change.

只要不生成新的物质，任何变化都是物理变化。(译为条件状语)

At the center of an atom there is a nucleus, around which electrons revolve.

原子的中心是原子核，电子围绕原子核旋转。(译为并列分句)

(2) 名词（代词） + of + which 的定语从句的翻译。

“名词（代词） + of + which” 结构的定语从句主要是非限制性定语从句，翻译时一般把 “of which” 翻译成 “其” 或 “它的”。例如：

There are 107 known elements, most of which are metals.

已知元素有 107 种，其中大部分是金属。

Perhaps light is some sort of electric wave, the nature of which we do not yet understand.

也许，光是某种电波，其性质我们尚不清楚。

有时也可重复 which 指代的先行词，将定语从句译为分句。例如：

This kind of chemicals, the evaporating temperature of which is low, must be kept at a temperature below zero Centigrade.

因为这种化学药品的蒸发温度很低，故其保存温度应在零摄氏度以下。

(3) 割裂式定语从句的翻译。

通常情况下，英语中的定语从句是紧随先行词之后的，但为了表达的需要，有时也会被其他成分分开，形成割裂式的修饰现象，这种情况在科技英语中尤为多见。翻译这种定语从句时，首先要根据句子的句法结构和逻辑关联，分析并判断出定语从句所修饰的先行词，然后再根据从句和先行词之间的逻辑关系，灵活采用适当的定语从句的翻译方法加以处理。例如：

In recent years ways have been developed by which air can be safely used over and over in space.（定语从句修饰 ways）

近年来，研发了一些能在太空中安全地反复利用空气的方法。

The forces acting on a given body which are exerted by other bodies are referred to as external forces.（定语从句修饰 forces）

由别的物体作用于某一物体上的力称为外力。

These limitations do not apply to operations like boring which are performed with simple point cutter.（定语从句修饰 operations）

这些限制不适用于一些像镗削那样的加工，因为这些加工是使用单刃刀具进行的。

（4）由 as 引导的定语从句。

由 as 引导的非限制性定语从句通常修饰前面的整个主句或主句的一部分内容，引导词 as 在定语从句中作主语，从句的位置可放在主句之前，也可以像其他非限制性定语从句一样后置，其功能主要是对主句作附加说明，或表示说话人的看法或态度，常常可以译为“正如……那样”或“像……”。例如：

Negative feedback, as is described here, is most widely applied in the automatic controls of various kinds of mechanism.

负反馈，正如这里描述的那样，非常广泛地应用在各种机械的自动控制装置中。

As will be explained in chapter 5, the body has natural defenses against these organisms.

正如在第 5 章将要讲到的那样，人体具有抵抗这些生物体的天然防御机制。

Almost all metals are good conductors of electricity as has been stated above.

如上所述，几乎所有的金属都是电的良导体。

科技英语中，许多由 as 引导的定语从句使用频繁，已形成了固定的表达，可作为固定搭配记忆和掌握。例如：

as is described above 如上所述

as is stated above 如上所示

as is explained before 如前所释

as is indicated in...如在……中所指出的那样

as have been already mentioned 正如前面所提到（讲过）的那样

as is pointed out many times 正如多次指出的那样

as is often said 正如通常所说的那样

as is known to all 众所周知

as might have been expected 正如所预料的那样

5.5 定语从句的其他翻译方法

英语定语从句的形式多样，用法多变，在科技英语中尤为如此，因此在翻译时可以根据具体情况做出适当的处理和调整，按照汉语的表达习惯和方式，最大程度地体现原文意思。

（1）融合法（Mixture）。

有些定语从句在翻译时，可以将定语部分和其修饰的先行词合并成主谓结构，把整个句子译成简单句，这种方法称为融合法。这种方法在处理“there be 句型 + 定语从句”的句子时特别适用。例如：

There are a number of factors that can affect the magnitude of resistance.

有许多因素影响电阻的大小。

There are bacteria that help plants grow, others that get rid of dead animals and plants by making them decay, and some that live in soil and make it better for growing crops.

有些细菌能帮助植物生长，另一些细菌则通过腐蚀来清除死去的动物和植物，还有一些细菌则生活在土壤里，使土壤更有利于种植庄稼。

There are some engine rooms on board sea-going ships where all the devices are automatically-controlled.

远洋轮船上有些机舱设备是自动控制的。

（2）定语从句译成汉语句子的（非状语）其他成分。

在翻译定语从句时，有时根据表达需要，可将定语从句，无论是限制性还是非限制性的，译成汉语的同位语、主语、宾语等其他成分，抑或转变为其他句式。这就需要译者从翻译实际需求出发，准确理解原文句意，应用适当翻译方法，灵活处理。例如：

The speed of sound in air at ordinary temperature is about 1,100 feet per second, which is about one mile in five seconds or about 700 miles per hour.

在常温下，声音在空气中的传播速度为每秒 1 100 英尺，即每 5 秒约 1 英里或每小时约 700 英里。(译为同位语)

As man destroys the forests, millions of species of plants and animals, the vast majority of which have not been discovered yet, will become extinct.

由于人类对森林的破坏，数百万种动植物（其中大多数还尚未被发现）将会灭绝。(加括号译为注释文字)

Recent Scientific innovations point to a future where tiny amounts of electricity will run superfast computers.

最近的科学创新表明，在未来，极少量的电能就足以运行超高速计算机。(译为宾语)

We live in an age where voice, data and video are just bits, ones and zeros.

在我们生活的时代，声音、数据、视频都是比特，即 1 或 0。(定语从句译为主句，主句译为状语)

Modern natural history is an exacting science whose practitioners must also cope and improvise in difficult field conditions.

现代自然历史是一门要求严苛的学科，不仅如此，其研究者还要想办法克服野外环境带来的困难。(译为并列句)

第 6 节　状语从句的翻译

状语从句在整个句子中起到状语的作用，按照其功能，可分为 9 大类型：时间状语从句、地点状语从句、原因状语从句、结果状语从句、条件状语从句、方式状语从句、目的状语从句、让步状语从句和比较状语从句。英语的状语从句和汉语的偏正结构中的状语成分相似，所以一般来说转换起来比较容易。但由于英语的状语从句在句中位置比较灵活，可放在主句前面，也可放在主句后面；而汉语中的状语成分的位置较为固定，一般位于句子前面，故在翻译时应注意语序的调整。另外，英语的状语从句一般由从属连接词或起连接作用的词组来引导，在翻译时一定要注意弄清引导词的含义，理清句子逻辑关系，区分各种不同的状语从句，从而提高翻译的准确性。

6.1　时间状语从句的翻译

英语中引导时间状语从句的连接词主要有：“when, whenever, as, before, after, until, till, since, while, once, each time”等，以及表示“一……就”的“as soon as, the moment, the instant, no sooner...than, hardly...when”等。翻译时间状语从句，主要采取的翻译方法有顺译法、倒译法和转译法。

（1）顺译法。

当英语的时间状语从句在主句之前，翻译时不用改变语序，直接将英语的时间状语从句译为汉语的时间状语即可。例如：

When small eddies develop in Jupiter's atmosphere, the Great Red Spot tends to suck them in.

当较小的漩涡在木星的大气层中形成时，木星大红斑往往会将其吞没。

Whenever the bullet left the gun, it produced the equal and opposite force known as “kick” or recoil.

每当子弹出镗时，就会产生一种大小相等、方向相反的力，即所谓的“反冲力”或后坐力。

Hardly had the operator pressed the button when all the electric machines began to work.

操作人员一按电钮，所有电机都开始运转。

有些时候，虽然英语中的时间状语从句在主句之后，但汉译时仍可按照原文语序进行翻译。这种情况常出现在 until 和 till 引导时间状语从句，且主句为肯定句的时候。例如：

This sound keeps going until it hits a new layer of mud or soil.

这种声音继续传播，直到碰到新的泥层或土层。

Boil the solution till it becomes free of H_2S.

加热该溶液直到没有硫化氢为止。

（2）倒译法。

英语的时间状语从句可以放在主句之后，汉译时，为了使译文符合汉语表达习惯，通常将状语成分放在句首。例如：

The spot of light describes a circular path when the lens assembly is rotated.

当透镜系统旋转时，光点画出一个圆形轨迹。

The productivity has been raised greatly since numerical control was adopted in machine tools.

自从机床采用数控方式以来，生产效率大大提高。

注意，由 until 和 till 引导时间状语从句，如果主句是否定句，汉译时通常采用倒译法将状语成分放在句首，译为“直到……才”。例如：

Crystal making did not become a serious business until 1902.

直到1902年，晶体制造才成为一个重要的行业。

The direct exploration of the upper atmosphere didn't become possible until the high-altitude rocket was invented.

直到发明了高空火箭后，人们才能直接探索上层大气。

（3）转译法。

有些时间状语从句虽然从语法上看用的是表示时间的关联词，但从逻辑上分析却相当于其他关系的状语从句（条件状语从句较为常见），在翻译时可以将这种时间状语从句转译成其他状语从句，同时还要注意语序的调整。例如：

A body at rest will not move till a force is exerted on it.

若没有外力作用，静止的物体不会发生位移。（转译为条件状语）

Researchers found that physicians who had computers ordered nine percent fewer tests when the software told them beforehand the patient's logistical probability of having an abnormal result.

研究人员发现，如果计算机软件事先把患者出现异常结果的可能性告诉医生，那么他们开出的检验项目可减少9%。（转译为条件状语）

When it has all these details, the central computer sends back to the driver via the buried electronic unit the best possible route.

中央计算机一旦获得所有这些详情，就能通过埋在道路下的电子装置将最佳路线提供给驾驶人员。（转译为条件状语）

Stress during labor comes from the periodic reduction in the oxygen supply when the pressure of the contraction stops blood flow through the placenta.

由于子宫收缩阻止血液通过胎盘，就会引起间歇性供氧减少，从而导致分娩应激反应。（转译为原因状语）

还有一种情况需要注意，在翻译由 when 引导的时间状语从句时，有时并不把 when 译为“当……”，而是译成“……之后”或“自从……”。例如：

When the thermometer becomes warmer the mercury expands, and the amount of expansion can be noted on a scale engraved on, or attached to, the stem.

温度计受热后，水银就会膨胀，其膨胀程度可以从刻在或附在管柱上的刻度读出。

Living standards have been found to improve substantially when electricity is introduced for such purposes as lighting, refrigeration, irrigation and communication.

自从使用电来照明、制冷、灌溉以及通信以来，人们的生活水准有了大幅提高。

6.2 地点状语从句的翻译

英语中引导地点状语从句的连接词较少，常用的有 where 和 wherever。地点状语从句翻译方法主要包括顺译法、倒译法和转译法。

（1）顺译法。

无论地点状语从句是位于主句之前还是主句之后，翻译时均有可能按照原文语序进行顺译。例如：

Where there is no need for electrical isolation an auto-transformer is sometimes employed.

在不需要电绝缘的地方，有时使用自耦变压器。

We use insulators to prevent electrical charges from going where they are not wanted.

使用绝缘体是为了防止电荷跑到不需要的地方去。

（2）倒译法。

当地点状语从句在主句之后，汉译时有时需要根据情况将汉译后的状语成分调整到句首。例如：

The current has the same value wherever in the circuit it is measured.

在电路的任何地方测电流，其大小都相同。

Lubricity is of primary concern where moving parts are in contact.

在运动部件相互接触之处，润滑是极为重要的。

（3）转译法。

有时地点状语从句在逻辑关系和意义上相当于其他状语从句，翻译时要根据具体情况将其转译为表达相应关系的汉语状语成分，并注意语序的调整。例如：

Where a vessel has vertical sides, the pressure on the bottom is equal to the height of the liquid times its density.

如果器壁垂直，则容器底部压强等于液体高度乘以液体的密度。（转译为条件状语）

The micrometer may be applied, where the meter is too large a unit to be used conveniently.

若用米作长度单位过大，则可改用微米。（转译为条件状语）

Where the circuit is complex, the test sequence needs to be computer controlled.

当遇到复杂的电路时，需要用计算机控制测试序列。（转译为时间状语）

At speed faster than the speed of sound, the plane leaves the sound waves behind where they cannot cause the plane any trouble.

飞机以超音速飞行时，把声波抛在后边，这样声波就不会给飞机造成影响。（转译为结果状语）

6.3 原因状语从句的翻译

英语原因状语从句的连接词主要有“because, since, as, for, now that, in that, for the reason that, considering that”等。英语原因状语从句的位置比较灵活，可放在主句之前，也可放在主句之后，而汉语的原因状语成分通常位于句首，因此在翻译时应注意语序的调整。

（1）顺译法。

原因状语从句位于主句之前时，通常采取顺译的方式进行翻译。

Since sound travels through air, there is no sound without it.

由于声音是通过空气传播的，所以没有空气就没有声音。

Because two or more steps were involved, the processes came to be known as indirect processes.

因为这些冶炼法需要两步或多步完成，因此被称为间接法。

有些时候，某些原因状语从句位于主句之后，但其意义在于对前面的结果进行补充说明，这时在汉译时无须调整语序，仍然可以采取顺译的方式。例如：

The material first used was copper for the reason that it is easily obtained in its pure state.

最先使用的材料是铜，因为易于制取纯铜。

Among the advantages of this system is the almost complete elimination of the need for pesticides and herbicides, as the controlled conditions reduce the possibility of pests and diseases.

这一系统的一个优点就是，几乎不需要任何杀虫剂和除草剂，因为受控条件减少了病虫害产生的可能性。

（2）倒译法。

当原因状语从句位于主句之后，而根据汉语表达习惯，汉译后的原因状语成分需要出现在句首，这时通常采取倒译方式进行翻译。例如：

Some sulphur dioxide is liberated when coal, heavy oil and gas burn, because they all contain sulphur compounds.

因为煤、重油和煤气中都含有硫化物，所以它们燃烧时会释放出二氧化硫。

The various types of single-phase a. c. motors and universal motors are used very little in industrial applications, since polyphaser a. c. or d. c. power is generally available.

因为通常有多相交流电和直流电可用，因此在工业上很少使用各种类型的单相交流电动机和交直流两用电动机。

He stuck to his opinion because he was convinced of the accuracy of this fact.

因为他深信这一事实正确可靠，所以坚持己见。

6.4 结果状语从句的翻译

英语结果状语从句的连接词主要有“so that, so…that, to the extent that, in such a way that, with the result that”等。这类状语从句在英汉两种语言中通常都放在主句之后，故多数情况下采用顺译的方式即可。表示结果关系的连接词通常译为汉语的“因而”“致使”“如此……以致”等。例如：

The sun is so much bigger than the earth that it would take over a million earths to fill a ball as big as the sun.

太阳比地球大多了，所以得有一百多万个地球才能填满像太阳那么大的球体。

In the magnet the atoms are lined up in such a way that their electrons are circling in the same direction.

在磁体中，原子排列的方式使电子沿同一方向作圆周运动。

A much higher temperature is required so that we could change iron from a solid state into liquid.

需要更高的温度，才能把铁从固态变为液态。

6.5 条件状语从句的翻译

英语中引导条件状语从句的连接词主要有“if, as (so) long as, unless, provided (providing) that, suppose (supposing) that, assume (assuming) that, in case, on the condition that, in the event that, unless”等。条件状语从句在翻译时可采用顺译法、倒译法和转译法。根据具体情况，英语连接词可用汉语表示条件关系的关联词“如果”“若”“只要……就”“除非……才”等进行翻译。

(1) 顺译法。

当英语条件状语从句在主句之前时，汉译时无须调整语序，直接采取顺译的方式。例如：

If the medium is a solid—in which case the electrons are more tightly packed—the electron flow will be slower.

如果介质是固态，电子则聚集较紧密，这时电子的流动就会相对缓慢。

Providing that too much current flowed through an ammeter it would cause the voltage to fall when it was connected, and the measured voltage would be wrong.

假如通过安培计的电流太多，就会使所连安培计电压下降，从而导致所测电压值数据错误。

Were there no electric pressure in a semiconductor the electron flow would not take place in it.

若半导体没有电压，其内部就不会产生电流。

注意：

此句是一种特殊的条件句，即虚拟语气的用法。这类句子省略 if 引导词，将 were, had, should 等词提到句首。

另外，英语中也有个别的条件状语从句虽位于主句之后，但由于其表示的是对主句的补充说明，在翻译时也可按照原句序，将汉译后的条件状语成分放在句尾处理。例如：

The words "velocity" and "speed" are considered as synonyms unless they are used in technical books.

"velocity"和"speed"这两个表示"速度"之意的词可视作同义词，除非是在科技书籍中使用。

(2) 倒译法。

英语中的条件状语从句位置灵活，除了可放在主句之前，还可以放在主句之后，而汉语的条件状语成分通常在句首出现，这时就需要采用倒译的方式调整语序。例如：

There is no charge in the motion of a body unless a resultant force is acting upon it.

除非有一个合力作用于物体，否则物体的运动状态不会发生改变。

All gases at 0 ℃ are found to gain 1 part in 273 of their volume when heated to 1 ℃, so long as the pressure is maintained constant.

只要压力保持不变，所有气体从 0 ℃加热到 1 ℃，体积就增大 1/273。

The design is likely to be accepted on the condition that the cost is reasonable.

假如成本不高，这项设计很可能会被接受。

(3) 转译法。

有些时候，if 引导的条件状语从句在翻译时并不一定都很机械地翻译成"如果……"的汉语句子，而是可以译为"当……时"的表示时间关系的句子。例如：

Electricity would be of very little service if we were obliged to depend on the momentary flow.

当我们需要依靠瞬间电流时，电就没有多大用处了。

If the distance from the generator to the load is considerable, it may be desirable to install transformers at the generator and at the load end, and to transmit the power over a high-voltage line.

当发电机离负载有相当距离时，最理想的方法是在发电机和负载端安装变压器，并采用高压线路输电。

If the admissible gripping force is exceeded, a hydraulic overload protection is operated.

当加紧力超过了允许值，液压超负荷防护装置就会起作用。

6.6 方式状语从句的翻译

英语的方式状语从句的连接词有"as, just as, as if, as though, in a manner that, in this (such) way that”等，汉语通常译为“正如……”“就像……”。“as if, as though”引导的状语从句多用虚拟语气，表示与事实相反，汉语常译为“好像……似的”“犹如……一样”。英语的方式状语从句和汉语的方式状语在句中的位置均较为灵活，可根据具体情况灵活处理。常用的方式状语从句的翻译方法主要有顺译法和变序译法（Rearranging the sentence order）两种。

（1）顺译法。

Just as sound waves do, light waves differ in frequency.

同声波一样，光波也有不同的频率。

Most plants need sunlight just as they need water.

多数植物需要阳光，就像它们需要水一样。

When the computer finds the closest match, it encodes the character in memory and displays it on the screen as if it had been typed.

一旦找到最匹配的字符，计算机就在内存中将该字符编码，并在屏幕上显示，如同从键盘上敲入该字符似的。

（2）变序译法。

在翻译方式状语从句时，有时为了符合汉语的表达习惯，可将英语中的方式状语从句插入汉语的主谓之间，将原从句译为汉语的简单句。例如：

Just as all living things need air, water and sunlight, so plants need them.

植物就像所有生物一样也需要空气、水和阳光。

Just as steel is an important material, aluminum alloy is also an important one.

铝合金和钢一样，也是一种重要材料。

The primitive man lived in the damp blackness of vast forests, as the pygmies of African do to this day.

原始人就像现在的非洲矮人一样，住在大森林中潮湿阴暗的地方。

6.7 目的状语从句的翻译

目的状语从句的连接词常用的有“so that, in order that, for the purpose that, for fear that, lest, in case, to the end that”等。需要注意的是 so that 既可以引导结果状语从句，也可以引导目的状语从句，二者之间的区别在于：结果状语从句表示一种事实，通常译为“以至于”“因此”；而目的状语从句表示某种可能性或要达到的愿望，通常译为“为了……”“要使……”，且英文谓语中也往往有 may, might, could, should 等情态动词。另外，结果状语从句通常在主句之后，而目的状语从句的位置在主句之前和之后都可以。

此外，for fear that, lest, in case 引导的目的状语从句要用虚拟语气，一般在主句之后出现。翻译目的状语从句一般采用顺译的方式，即遵循原句语序，不做改变。要注意的是，当目的状语从句在主句之前时，译为“为了……”“要使……”等，而当目的状语从句在主句之后时，译为“以便”“以免”“以防”等。例如：

For the purpose that useful work from the chemical energy stored in fuels might be produced, it is necessary first to convert the chemical energy into heat by combustion.

为了可以将储存于燃料中的化学能转变为有用功，首先需要通过燃烧将化学能转变为热能。

In order that electrons may be accelerated in an applied field, they must be able to move to higher energy levels.

为了使电子能在外加场中加速，电子应能够移到更高的能级才行。

Iron products are often coated lest they should rust.

铁制品常常涂上保护层，以免生锈。

In the process of preparing for possible conflict, the operational commander must plan the execution of his campaign so that all operational level activities or operational functions can be synchronized.

在防范可能出现的冲突的过程中，作战指挥官必须筹划其所指挥的战役的实施步骤，以便所有作战层次的行动或作战功能能够实现同步。

6.8 让步状语从句的翻译

让步状语从句常用的连接词有“although, though, even though, even if, while”等，通常译为汉语的“虽然”“即使”“尽管”等；还可用“no matter how（however）, no matter what（whatever）, no matter when（whenever）, no matter where（wherever）”等连接词连接，通常译为汉语的“不论”“无论”“不管”等；也可用“as”连接，这时as引导的让步状语从句需要用倒装句式。英语中让步状语从句可置于主句之前，也可置于主句之后，而汉语让步关系的状语成分通常在句首出现，翻译时应注意语序的调整。

（1）顺译法。

若英语让步状语从句在主句之前，或有时插入在句中，翻译时无须改变语序，直接采用顺译的方式进行翻译。例如：

Although these particles are very light, their energies are considerable because of their high speeds.

虽然这些粒子质量很小，但由于（其运动）速度极快，因而能量很大。

Even though it is not so strong as the earth's, the moon's gravity does something to the earth.

尽管月球的引力不像地球引力那么大，但它对地球仍会产生影响。

注意：

上句中让步状语从句的主语和主句的主语指代的是相同事物，这时从句中的主语可用代词替代。这种情况在汉译时需要注意进行两个主语的换位，即，将从句中作主语的代词替换成名词，而把主句中作主语的名词替换成代词。

No matter what the shape of a magnet may be, it can attract iron and steel.

不论磁铁形状如何，它都能够吸引钢和铁。

All matter, whether it is solid, liquid or gas, is made of atoms.

所有物质，无论是固体、液体还是气体，都是由原子构成的。

Complicated as a modern machine is, it is essentially a combination of simple machines.

现代机器虽然复杂，但实质上不过是由许多简单机械组合而成的。

（2）倒译法。

当让步状语从句位于主句之后，汉译时通常用倒译的方式进行语序的调整。例如：

All electronic computers consist of five units although they are of different kinds.

尽管种类各异，但电子计算机都是由五大部分组成的。

Concrete block walls will be provided wherever it is required for fire protection.

凡是需要防火墙的地方都将采用混凝土砖块砌墙。

Stars are in reality enormous bodies while they look like mere specks.

虽然星星看起来只是些光点，但实际上它们都是极其巨大的星球。

6.9 比较状语从句的翻译

英语中的比较状语从句通常由“as…as（像……一样），not so（as）…as（不像……一样），more…than（比），no more…than（并不比），not more…than（不如，比不上），the more…the more（越……越）”等引导。英语的比较状语从句经常以省略句的形式出现，即从句中只出现和主句进行比较的部分，而省略和主句意义相同或重复的部分。翻译比较状语从句通常采用顺译法进行。需要注意的是，汉语表示比较的方式比较多样，翻译时需根据汉语的表达习惯灵活处理。例如：

Electronic circuits work a thousand times more rapidly than nerve cells do in the human brain.

电子电路的工作速度是人脑神经细胞的 1 000 倍。

The faster the gas rushes out, the faster the rocket moves.

气体喷出速度越快，火箭运行速度也就越快。

These bearings are not as well standardized as the rolling contact bearings.

这些轴承的标准化程度不及滚动轴承。

第 7 节 否定句的翻译

每种语言都有其表达肯定和否定的语法结构和方式，英语和汉语也不例外。本节主要讨论科技英语中各种否定句的翻译方法。由于英汉两种语言不同特点和表达习惯，英语和汉语用于表达否定概念时所使用的词汇、语法结构和逻辑结构等都存在着很大差别。英语中表达否定的形式十分多样，有全部否定、部分否定、准否定、意义否定和双重否定等多种句式；

而在汉语中表达否定的形式相对简单，因此翻译以上否定句型和结构存在一定难度。另外，在翻译英语的否定句时，还会遇见两种较为特殊的情况，即否定的转移和否定的延续，如果不能清楚地了解原句的否定意味，就容易产生误译。本节主要讨论如何正确理解科技英语中用于表达否定的词汇和句法结构，帮助大家准确有效地翻译各种类型的否定句式。

7.1 全部否定句

全部否定即否定整句的意思，英语中常常通过“not，no，none，never，neither，nothing”等否定词来实现。这种表达否定的句式同汉语的否定句比较对等，翻译的难度不大。例如：

At least nowadays, there is no way to harness the energy of fusion.

至少目前还没办法利用核聚变。

Hardened steel and brittle materials such as glass and ceramics are not normally amenable to diamond turning.

一般情况下，淬硬钢材或玻璃、陶瓷等脆性材料不宜使用金刚石切削。

A proton has a positive charge and electron has a negative charge, but neutron has neither.

质子带正电荷，电子带负电荷，中子不带任何电荷。

7.2 部分否定句

部分否定句，即否定句子整体中的一部分。这种句式是由否定词“not”搭配“all，everything，both，many，each”等表示全体意义的不定代词，或搭配“always，often，altogether，entirely，wholly”等副词构成的。翻译这种句式务必注意不要译成“一切……都不”的表达全部否定的句子，而要译为“不都是”“并非都是”“不总是”等表达部分否定的汉语句式。例如：

All forms of matter do not have the same properties.

并非所有形式的物质都具有相同的特性。

Not all of the heat supplied to the engine is converted into useful work.

并非供给热机的所有热量都转变为有用的功。

注意：

部分否定句中的否定词 not 可以放在句中，也可放在句首。

This plant does not always make such machine tools.

这个工厂并不只是制造这样的机床。

请注意，“表示全体意义的不定代词＋肯定式谓语＋含否定意义的单词（常见带否定含义前缀的单词）”这种否定句式表示的是全部否定，而非部分否定。翻译时，请务必注意辨析，准确把握句意，避免误译。例如：

All germs are invisible to the naked eye.
一切细菌都是肉眼看不见的。
Every design made by her is impossible of execution.
她所做的一切设计都是不能实现的。
Both data are incomplete.
两个数据都不完整。
In practice, error sometimes always seems unavoidable.
在实践中，差错有时似乎总是不可避免的。

7.3 准否定

英语中有一类否定句虽然没有完全否定，但肯定的含义极少，这种否定句叫作准否定，或几乎否定。表达准否定的句子中通常使用“few, little, hardly, scarcely, barely, seldom”等否定词，可译为汉语的“极少”“很少”“就没有”“几乎不”等。例如：

The earliest forms of jet-propulsion had little ability to function at rest, in view of the absence of any means of air-compression.

最早的喷气式发动机由于缺乏空气压缩手段，静止时几乎没有工作能力。

Barely any of our present batteries would be satisfactory enough to drive the electric train fast and at a reasonable cost.

目前的蓄电池几乎都不足以保证电气火车能够既经济又高速地运行。

Mercury, so small and close to the sun that its gases were quickly lost to space, is nearly airless.

水星几乎没有空气，因为它体积小，并且离太阳很近，所以它的气体很快就散失在太空中。

7.4 意义否定

英语中有些句子虽然语法结构是肯定形式，但是其内容却包含有否定的意义，这种情况称之为意义否定或内容否定。这类否定句中往往含有带否定意义的词或词组，如“absence, ignorance, lack, fail, neglect, far from, instead of, rather than”等。另外，英语中还有许多带有否定前缀和否定后缀的词汇同样具有否定意义，如 non-linear（非线性），unbalance（不平衡），impossible（不可能的），irreversible（不可逆的）等以及 useless（没有用的）等。翻译时遇到以上具有否定意义的词汇，一般添加汉语的“不”“没有”等否定词予以体现。例如：

In the muddy waters of South American rivers these fishes' eyes are of little use to them; instead of eyes they use extremely accurate electric sense organs.

在南美洲浑浊的河水里，鱼的眼睛几乎没有多少用处；这些鱼不靠眼睛，而是使用极为精准的电感器官。

In the absence of force, a body will either remain at rest or continue to move with constant speed in a straight line.

在没有外力的情况下，物体或保持静止，或继续做匀速直线运动。

Such very close unions are known as compounds rather than mixture.

这种非常紧密的结合物叫作化合物，而不是混合物。

A machine's information would be useless unless the programmer knew how to instruct it in what human beings refer to as commonsense reasoning.

要是程序员不懂得按照人类所谓的常识推理对计算机下达指令，电脑的信息还是没有用的。

7.5 双重否定

双重否定即否定之否定，这种否定句式虽然在形式上是否定结构，但实际却表达肯定含义。英语中的双重否定通常是由"no，not，never，nothing，nobody"等否定词和其他含有否定意义的词或词组连用而构成的。翻译这类双重否定句，可以直接译为汉语的双重否定句，也可以译为肯定句。

（1）译为双重否定句。

If it were not enough acceleration, the earth satellite would not get into space.

如果没有足够的加速度，地球卫星就无法进入太空。

In the absence of gravity, we could do nothing.

要是没有地心引力的话，我们什么也做不成。

But for substances, there would be nothing in the world.

若没有物质，世界上就什么东西也不存在了。

（2）译为肯定句。

Despite its many advantages, wood is not without its drawbacks as a fuel.

尽管木材有许多优点，但作为燃料还是有缺点存在的。

Friction is not always undesirable.

摩擦力有时也是必要的。

Heat can never be converted into a certain energy without something lost.

每当热能转换为某种能量时，总会有能量损耗。

7.6 否定的转移

在科技英语翻译中，常常会遇到这样的问题，即英语中的否定词在句中究竟是否定的哪个成分？有时，在形式上否定的是主语，但实际意义上却否定的是谓语；或者在形式上否定的是状语，但实际意义上却否定的是谓语；还有一些情况形式上否定的是主句，意义上否定的是从句。这类现象称为否定的转移（Transferred Negation）。汉语在表达上述否定意味时不存在这种现象，因此，遇到包含否定转移现象的句子，在英译汉的过程中，译者务必要注意这种英汉否定表述方式的不同，在正确理解原文的基础上，按照汉语的表达习惯，将否定的位置加以转换。

（1）一般的否定转移。

①某些英语简单句中会出现这类否定位置的转移现象。例如：

No defects have been found in these circuits.

这些电路中没有发现任何缺陷。(英语否定主语，汉语否定谓语)

We know of no effective way to store the Sun's heat.

我们不知道贮藏太阳热量的有效方法。(英语否定宾语，汉语否定谓语)

②英语中有些表示否定意义的介词短语如“under no circumstances, on no conditions, in no circumstances”等，在汉译时通常要将否定介词宾语（或否定状语）转译为否定谓语。例如：

Under no circumstances is aluminum found free in nature.

在自然界，任何情况下都不存在处于游离状态的铝。

On no conditions can energy be created or destroyed.

在任何情况下，能量既不能凭空产生也不会凭空消失。

③英语以动词“think, believe, expect, suppose, hope, imagine, anticipate”等作为谓语动词且后接不定式短语的否定句，原文中否定谓语动词，汉译时否定不定式短语。例如：

Liquids, except for liquid metal such as mercury, are not considered to be a good conductor of heat.

除了液态金属如水银外，其他液体则被认为是不良导热体。

For many years the atom was not believed to be divisible.

过去多年来人们一直认为原子是不可分割的。

（2）主句否定转移为从句的否定。

①碰到 not…because 句型，翻译时一定要注意正确理解句意，把握否定的位置，否则很可能发生误译。例如：

The engine didn't stop because the fuel was finished.

引擎并不是因为燃料耗尽才停止转动的。

The object did not move because it was pushed.

该物体并非由于受到推力作用而移动。

②在由“think, believe, expect, suppose, hope, imagine, anticipate”等这类动词后接宾语从句的否定句中，若否定词否定的是主句的谓语动词，翻译时应转换为否定宾语从句中的谓语。例如：

They didn't think that a death-ray could be produced.

他们认为不可能产生死光。

③由 as 引导的定语从句和方式状语从句本不难理解，但是，当这种复合句的主句为否定句，且 as 作为“像……一样”理解时，情况比较复杂，容易引起误解。多数情况下，可以把主句的否定含义转移到 as 引导的从句上来，通常可以译为“与……相反”，或译成“主句主语 + 并不像……那样”。例如：

Spiders are not insects, as many people think, nor even related to them.
与多数人所想的相反，蜘蛛并不属于昆虫类，甚至与昆虫没有近亲关系。

The earth does not move round in the empty space as it was once thought to be.
地球并不像人们曾经认为的那样是在空无一物的空间中运转的。
或：与人们曾经认为的相反，地球并不是在空无一物的空间中运转的。

注意：
当 as 引导的从句位于句首时，汉译时不用进行否定的转移。

As is known to all, man can't live without air.
众所周知，没有空气，人就不能生存。

7.7 否定的延续

在科技英语中会碰到这样的情况：在 and 或 or 连接的并列句中，前一分句谓语有否定词，后一分句谓语没有否定词。遇到这种情况，是只翻译前句的否定，还是要将否定延续到后一句，通常要根据各分句中谓语是否带有情态动词或助动词而定。

(1) 前一个分句谓语有情态动词、助动词，而后一个分句没有情态动词、助动词，需要进行否定延续。例如：

Carbon dioxide does not burn and support combustion.
二氧化碳既不能燃烧，也不助燃。

(2) 前一个分句谓语有情态动词、助动词，而后一个分句也有情态动词、助动词，不需要进行否定延续。例如：

Thus when coal is piled too deeply, heat formed by slow oxidation of coal within the pile cannot find a way out and may cause the pile to catch fire of itself.
因此，当煤堆得太厚，煤在煤堆里缓慢氧化产生的热无法散出，就可能引起煤堆自燃。

第 8 节 复杂长句的翻译

英语句子结构为形合式结构，也就是说英语可以借助各种连接手段、各种短语、不同从句以及多种语法手段对句子结构加以扩展和组合，形成层次繁杂的复杂长句。科技英语是用来陈述自然界、科技界所发生或出现的事情，描述其规律、特点、过程等的语言。为了实现客观准确、逻辑紧密、结构严谨的科技语言风格，科技英语中大量使用复杂长句，这是科技英语的一大语言特色。而汉语为意合式结构，句子少用甚至不用连接词语，句子简短明快，语段结构流散，但语义层次分明，句子之间的关系主要由上下文的语义来连贯。由于英汉两种语言结构的巨大差异，翻译长难复杂句是翻译工作者的一大难题。翻译科技英语复杂长句，一般要将长句拆成汉语短句，按照汉语的表达习惯和逻辑层次，重新排列顺序，组织成内容准确、逻辑分明、重点突出、通顺正确的译文。

进行复杂长句翻译时，首先要通读全句，分清句法结构。先分析句子是简单句、并列句还是复合句。如为简单句，则应先分析出主语、谓语、宾语、表语（主要成分），再分析出定语、状语等修饰语成分（次要成分），特别注意非谓语形式（不定式短语、分词短语、动名词短语）所作的修饰语；然后要弄清主要成分和次要成分之间的关系；同时注意时态、语气和语态等。如为复合句，则应先找出主句，再确定从句及其性质；对于各从句（定语从句、状语从句及各种名词性从句）则分别按简单句分析。

理解语法关系之后要领会句子的要旨和逻辑关系，分清重心，突出重点。英语句子习惯先传递主要信息，利用各种连接手段，再补充各种次要信息。而汉语句子多用语序表达逻辑层次，一般是由小到大、先因后果，重心在后。因此，汉译过程中要注意表达方式和表达习惯的转换。

在译文表达阶段，利用我们前面讨论过的有关词、短语、各种句式等的翻译方法，逐层翻译长句，再按照汉语的特点和表达方式重新组句，并进行加工润色。译文可以不拘泥于原文的形式，以正确表达原文内容和意义为最终目标。

翻译科技英语复杂长句可采用的方法主要有：顺译法、倒译法、分译法和综合译法四种。

8.1 顺译法

当英语复杂长句的表达顺序与汉语的表达顺序基本一致，可以基本维持原文语序和语法结构，采用顺译法进行翻译。所谓基本维持原文语序的意思并不是要求每一个词、每一个短语都完全按照原文语序，实际上完全和绝对的一一对应是极为少见的，这一点需要注意。例如：

Nations will usually produce and export those goods in which they have the greatest comparative advantage, and import those items in which they have the least comparative advantage.

分析：

本句是由 and 连接的并列句，前后两个并列分句中各包含一个由 in which 引导的定语从句，句子结构清晰明了，前后对比关系明确。该句表达顺序和汉语基本一致，只需将两处定语从句调整为汉语的前置定语，其他部分照译即可。

译文：

各国通常生产和出口那些最具相对优势的产品，而进口那些最不具相对优势的产品。

Chemists study the structure of food, timber, metals, drugs, petroleum and everything else we use to find out how the atoms are arranged in molecules, what shape the molecules have, what forces make the molecules arrange themselves, into crystals, and how these crystals arrange themselves into useful substances.

分析：

通过分析可以看出，这个复杂长句的句子主干是 Chemists study…to find out…，复杂之处在于 find out 后面跟了四个由 how 和 what 引导的宾语从句。该句表达顺序和汉语基本一致，故采用顺译法进行翻译。

译文：

化学家们研究食物、木材、金属、药品、石油以及其他我们所用之物的化学结构，以期发现原子在分子中是如何排列的，分子具有什么样的形状，是什么使分子排列成为晶体，而晶体又是如何排列成为有用的物质的。

However, even if prediction becomes possible, people who live in areas where earthquakes are a common occurrence will still have to do their best to prevent disasters by building structures that are resistant to ground movement and by being personally prepared.

分析：

本句是由 even if 引导的让步状语从句，从句部分较为简单。主句较为复杂，句子主干是 People will have to do their best to prevent disasters. 。在主干的基础上添加了若干个修饰成分。首先主语 people 后跟由 who 引导的定语从句，而该定语从句中又套了一个由 where 引导的定语从句修饰 area 一词；句子末尾处介词 by 引导的两个方式状语修饰前面的谓语 prevent disasters，需要留意的是两个 by 后面跟的都是动名词结构，其中一处还有一个由 that 引导的定语从句修饰 structures. 该复杂长句状语从句在前，主句在后的表达方式同汉语一致，总体上可采用顺译法进行翻译。

译文：

然而，即使地震可以预测，居住在地震频发区的人们还是应尽力预防灾难，办法是建造能够抗震的房屋，同时做好个人准备。

8.2 倒译法

汉语在叙述时一般是“先发生的事情先说，后发生的后说”“先原因，后结果”“先次要，后主要”。而英语有许多时候与汉语正好相反，会把主句放在句首，分析或说明部分置后。因此，当科技英语长句的叙述层次与汉语逻辑相反时，翻译要按照汉语的习惯从原文的后面译起，这种方法叫作倒译法。例如：

Aluminum remained unknown until the nineteenth century, because nowhere in nature is it found free, owing to its always being combined with other elements, most commonly with oxygen, for which it has strong affinity.

分析：

本句主干关系是由 because 引导的原因状语从句。其中从句中又包含了“owing

to…”分词结构构成的原因状语和 “for which…”引导的定语从句。翻译时，为了符合汉语表达习惯，我们可以采取倒译法，先翻译后面的原因状语从句，再翻译主句。另外，由于原因状语从句中还内嵌了因果关系，翻译时，同样先说原因，再说结论。这样，句子内容环环相扣，条例清晰，层次分明。

译文：

铝总是跟其他元素相结合，最普遍的是跟氧元素结合，因为铝和氧有很强的亲和力。由于这个原因，在自然界找不到游离状态的铝，所以铝直到 19 世纪才得以发现。

Being able to receive information from any of a large number of separate places, carry out the necessary calculations and give the answer or order to one or more of the same number of places scattered around a plant in a minute or two, or even in a few seconds, computers are ideal for automatic control in process industry.

分析：

本句是一个简单句。句子类型虽为简单句，但句子结构却很复杂。句子主干部分很简单，即为 computers are ideal for automatic control in process industry. 。而较为复杂的部分是前面 being able to…的分词结构作状语的结构，而且 to 后面并列连接了 receive, carry out, give 三个动词结构，还有一个 scattered 的过去分词作定语的结构。虽然一般情况，汉语表达是先说原因，再说结果，但这句由于原因的部分内容较为复杂，如果放在句前，会显得“头重脚轻”，因此在翻译本句时，我们可以使用倒译法，将句子译成汉语的总分结构，先说后面的主体核心思想，再从细节方面进行说明，阐述具体的原因，对前面的理念予以支持。

译文：

利用计算机在流程工业上进行自动化控制是最为理想的。这是因为，计算机可以获得分散在工厂周围的许多点的信息，并进行必要的运算，然后在一两分钟，甚至是几秒钟内，给其中的一个点或多个点做出回答或发出指令。

Scientists are learning a great deal about how the large plates in the earth's crust move, the stresses between plates, how earthquakes work, and the general probability that given place will have an earthquake, although they still cannot predict earthquakes.

分析：

本句是由 although 引导的让步状语从句。从句部分很简单，主句部分较为复杂，在“learn a great deal about”后面接了四个宾语，其中有句子，也有短语（也可看作是省略句），最后一处宾语 probability 后还跟有一个同位语从句。翻译时应按照汉语表达习惯，采用倒译法，先翻译后面的状语从句的部分，再翻译前面的主句。此外，在翻译“about”后面的四个宾语时，由于内容较多，也可以采取倒译的方式，先翻译宾语部分，再翻译谓语，这样逻辑概念会更为清晰。

译文：

尽管科学家仍无法预测地震，但对地壳中的大板块如何运动，板块间的压力如何，地震如何发生，某地区发生地震的一般概率是多少，科学家了解得越来越多。

8.3 分译法

有些科技英语复杂长句包含多层意思，且主句与从句或主句与修饰语之间的关系不是很紧密，如果翻译时沿用原句的结构，用一个句子对原文进行翻译，会引起句意混乱、逻辑不清等问题。对于这种长句，可以将原繁杂的长句拆分开，按照汉语多用短句的行文习惯，将长句中的从句或短语化为短句，分开进行叙述，为保证语意连贯，短句之间有时可适当增加一些关联词语。分译的目的是使句意表达清楚明了，逻辑关联性强，译文有整体感。例如：

Half-lives of different radioactive element vary from as many as 900 million years for one form of uranium, to a small fraction of a second for one form of polonium.

分析：

本句是一个简单句。句子主干可简化为 Half-lives of different radioactive element vary from…to…。本句的难点就在于 from 和 to 后面跟的宾语部分很复杂，如果按照原句结构翻译，会使译文表述不清，且不符合汉语表达习惯。通过分析，我们可以发现，本句包含两层意思：一是不同元素的半衰期不同；二是用具体例子说明不同元素的半衰期有很大差异。因此，在翻译本句时要进行句子结构拆分，这样一来，句子意思一目了然，结构清楚，也符合汉语总分式的表述方式。

译文：

不同的放射性元素其半衰期也不同，铀元素的一种同位素的半衰期长达9亿年，而钋元素的一种同位素的半衰期却短到几分之一秒。

This hope of “early discovery” of lung cancer followed by surgical cure, which currently seems to be the most effective form of therapy, is often thwarted by diverse biological behaviors in the rate and direction of the growth of cancer.

分析：

本句的主句部分是 This hope of “early discovery” of lung cancer followed by surgical cure is often thwarted by diverse biological behaviors in the rate and direction of the growth of cancer.，中间插入了一个由 which 引导的非限制性定语从句修饰 surgical cure 一词。如果按照原句的叙事方式进行翻译会使译文意思混乱不清，也无法抓住句子的要点。实际上，通过分析，我们可以发现本句包含了两层意思：一是人们希望能够早期发现肺癌，并通过相关医学手段进行有效治疗；二是由于某些特定原因，这种希望往往无法实现。因此，在翻译时，我们有必要将主句拆成两部分，先译谓语动词 is 之前的部分，即主语加定语从句的部分；再译后面的谓语部分，注意主语“hope”一词再次译出；by 一词连接的介词短语部分在这里译作原因状语。最后，前后两句之间需要添加表示转折的关联词，使句子浑然一体。

译文：

人们希望肺癌能够“早期发现”，随之进行外科治疗，因为外科治疗目前可能是最有效的办法。然而，由于肺癌在生长速度和生长方向等生物性特征上有很大差异，早期发现的希望往往落空。

The fact that present-day cancer rates do not exhibit any generalized increase (apart from the effects of smoking) does not, of course, guarantee that all, or nearly all, of the recently introduced new chemicals, pesticides, pollutants, and habits are harmless for, it may be decades before any cancer-causing effects may have become clearly evident.

分析：

这是一个结构颇为复杂的句子，句子的主干是 The fact...does not guarantee....第一处从句是 that 引导的同位语从句，解释和说明 fact 的内容；第二处从句是 guarantee 之后跟的由 that 引导的宾语从句；第三处是由 for 一词引出的原因状语从句；第四处是 before 引导的时间状语从句。翻译本句时可将原句拆成三个部分，即 does not 之前的主语部分、does not 之后的谓语动词 guarantee 加宾语从句部分以及最后的原因状语从句部分。先陈述："当今癌症发病率并无明显普遍上升趋势"，然后再进一步进行补充说明："这当然不足以说明最近采用的新型化学药物、杀虫剂、污染物和生活习惯等部分乃至全部均是无害的"，最后对前面这一观点进行解释和说明："以上致癌因素或许还要等几十年的时间才会显现出来"。

译文：

当今癌症发病率并无明显普遍上升趋势，但这当然不足以说明最近采用的新型化学药物、杀虫剂、污染物和生活习惯等部分乃至全部均是无害的，因为以上致癌因素或许还要等几十年的时间才会显现出来。

8.4 综合译法

在处理有些科技英语复杂长句时，有时单纯使用某一种翻译方法不能完全有效地完成翻译，这就要求我们对原文进行仔细推敲，兼顾上下文关系，按照叙述逻辑关系，主次分明地对全句做出综合处理，即在翻译时将顺译法、倒译法或分译法等方法综合运用，以把英语原文译为忠实通顺的汉语句子为最终目的。例如：

The phenomenon describes the way in which light physically scatters when it passes through particles in the earth's atmosphere that are 1/10 in diameter of the color of the light.

分析：

该句主句部分较为简单，即"The phenomenon describes the way."。句子的复杂之处在于环环相扣的从句结构，主要的从句关系是 in which 引导的定语从句修饰 the way, 定语从句中又包含了一个由 when 引导的时间状语从句修饰 light physically scatters, 最后还包含了一个由 that 引导的定语从句修饰 particles, 如何将复杂套接的从句部分翻译到位是我们需要解决的问题。翻译时，在综合考虑的前提下，首先用顺译加倒译的方式将"The phenomenon describes the way in which light physically scatters when it passes through particles in the earth's atmosphere."译为："这种现象说明了光线通过地球大气颗粒时的物理散射方式"。由于"that are 1/10 in diameter of the color of the light"这个定语从句只是对前面的先行词 particles 起到补充说明作用，所以对这部分采用分译的方法进行翻译。

译文：

这种现象说明了光线通过地球大气颗粒时的物理散射方式，大气微粒的直径为有色光直径的1/10。

The nutritional dilemma of persons with AIDS, or PWAS, is that they must eat abundantly yet they must also avoid food borne illness, and surmount the poor appetite and indigestion caused by their medications and by AIDS itself.

分析：

本句主体句子结构并不十分复杂，主体关系是 that 引导的表语从句，其中层次较为复杂的是表语从句中包含正反两层意思，由 yet 一词进行转折。yet 引出的转折部分又包含两部分概念，由 and 进行连接。翻译时，先采取分译法将主语的定语部分 persons with AIDS, or PWAS 单独译出：艾滋病患者又称 PWAS, 交代清楚事件对象，然后采取顺译法翻译"The nutritional dilemma is that they must eat abundantly.", 最后采取顺译加分译的方式，分层次翻译"yet they must also avoid food borne illness, and surmount the poor appetite and indigestion caused by their medications and by AIDS itself"。这样的处理方法使译文句子结构清楚，逻辑分明。

译文：

艾滋病患者又称 PWAS, 他们所遭遇的饮食困境表现为患者一方面必须大量进食，另一方面又不得不防范因饮食而导致的疾病，还要克服因药物治疗以及艾滋病本身而引发的食欲不振和消化不良。

The method normally employed for electrons to be produced in electron tubes is thermionic emission, in which advantage is taken of the fact that, if a solid body is heated sufficiently, some of the electrons that it contains will escape from its surface into the surrounding space.

分析：

句子的主句为"The method normally employed for electrons to be produced in electron tubes is thermionic emission.", 本句较为复杂的是主句后面跟随的非限制性定语从句中还包含了一个由 fact that 引出的同位语从句，同位语从句本身又是一个由 if 引导的状语从句。理解该句的核心要点在于能否洞悉"the method""thermionic emission"以及"the fact that 引出的同位语部分"实质上指的是一个概念——"热离子发射"。一旦理解了这一层内在关联，该句的翻译就迎刃而解了。翻译时，我们采取综合法，用倒译加分译将句尾同位语从句和主句中心词 thermionic emission 联合起来，译出"热离子发射"这一概念的定义内容，然后再译主句中的其他部分，即"热离子发射"的实际应用，并将 method 一词引申译为原理。这样的翻译，条理清楚，逻辑分明。

译文：

固体加热到足够温度时，它所含的电子就会有一部分离开表面飞逸到周围空间中去，这种现象叫热离子发射。电子管通常就是利用这一原理产生自由电子的。

第9节 练习

一、思考题

1. 英语和汉语在被动语态的使用上有哪些区别？在将英文被动句译成汉语时，可以采用哪些翻译方法？

2. 在翻译 and 连接的并列句时，除了将 and 译为汉语的“和”或“并且”外，还有没有其他可能出现的情况？

3. 当翻译 it 引导的形式主语句时，如果主句的时态是过去时，翻译时有什么需要特别注意的地方？

4. 同位语从句和定语从句有什么不同，同位语从句主要有几种翻译的方法？

5. 英语的定语从句和汉语的定语结构有什么不同？限制性定语从句和非限制性定语从句在翻译时有没有区别？

6. 英语的状语从句有多少种类型？将英语的状语从句转化为汉语的状语成分，有哪些地方需要特别注意？

7. 英语中的否定句都有哪些常见的形式？翻译英语否定句时最需要注意的是什么？

8. 科技英语中为什么经常使用复杂长句？翻译复杂长句的一般步骤是什么？翻译复杂长句的难点是什么？

二、翻译实践

1. 翻译下面的句子，注意被动语态的翻译。

(1) Since the first test of the atomic bomb the world has learnt the atom can be split and its power can be used.

(2) Each of the body systems is regulated in some way by some part of the endocrine system.

(3) Three groups of biologists, working in the years since the Renaissance, will be studied here.

(4) This suggests that cosmic rays do not come from all over the Universe, but are produced within galaxies.

(5) Should the blood become either too diluted or too viscous, the kidneys retain or excrete surplus water until the proper balance is reached.

(6) The volume is not measured in square millimeters. It is measured in cubic millimeters.

(7) On the TV screen the skylab could be seen orbiting round the earth in outer space.

(8) The surface would have to be protected from too frequent, or too intense, or too careless visiting.

(9) The deficiency of such observations will be eliminated by the installation of more ocean bottom seismographs.

(10) The increased efficiency in military operations brought about by advancements in information technology is considered by some to indicate another revolution in military affairs.

(11) Computer can deal with different kinds of problems if they are given the right instructions

for what to do.

(12) The rusting of iron is one example of corrosion, which may be described as the destructive chemical attack on a metal by media with which it comes in contact, such as moisture, air and water.

(13) We have already established that light may be treated as a transverse electromagnetic wave.

(14) Although the computer was devised mainly in response to scientific needs, the requirements of business and government data handling have been a major stimulant to the further development of this machinery.

(15) Almost everyone is involved with design in one way or another, even in daily living, because problems are posed and situations arise which must be solved.

(16) Other branches of mathematics such as algebra and geometry are also extensively used in many sciences and even in some areas of philosophy.

(17) Parts of numbers smaller than 1 are sometimes expressed in terms of fractions, but in scientific usage they are given as decimals.

(18) Much of this diversity is found in the world's tropical areas, particularly in the forest regions.

(19) A knowledge of statistics is required by every type of scientists for the analysis of data.

(20) In the binary scale, 2 is expressed as 010, 3 is given as 011 and 4 is represented as 100, etc.

2. 翻译下面的句子，注意连词“and”的翻译。

(1) The causes of accidents are extensively investigated, and it is usually found that a sequence of events precedes a catastrophe.

(2) Heat can also come from a forest fire and, in much smaller quantities, from the warm body of a mouse.

(3) Human error is often cited as the most frequent cause of accidents and this is rarely a simple matter.

(4) Thus, biological diversity helps prevent extinction of species and helps preserve the balance of nature.

(5) Keep your face to the sunshine, and you cannot see your shadow.

(6) The small three-person submarine is less than eight meters long, and it can dive almost four kilometers under the ocean.

(7) The phone of the future will be more mobile, do a lot of different tasks and be part of a complex, far-reaching information.

(8) Lead lacks tensile strength and it cannot be draw out in the form of fine wire.

(9) A black hole exerts a strong gravitational pull and it has no matter.

(10) Open the key and an induced current in the opposite direction will be obtained.

3. 翻译下面的句子，注意“it”作形式主语的句式的翻译。

(1) It is said that mathematics is the base of all other sciences, and that arithmetic, the

science of numbers, is the base of mathematics.

(2) It is reported that the National Security Agency is the largest employer of mathematicians in the United States.

(3) It could be said that mathematics is actually saving lives here because of the precision and accuracy made possible.

(4) It is important that the material always is dehumidified.

(5) It was tragic that nuclear fission was first developed for the production of atomic bomb.

(6) It has been demonstrated that electrons have magnetic lines of force around them.

(7) It should be realized that magnetic forces and electric forces are not the same.

(8) It was thought at one time that compound of carbon were only produced in living organisms.

(9) It has been taken notice of by foundry businesses all over the world that this technology has the advantages of great moulding speed, high production efficiency, fine technological performances, excellent overall economic results and good labor conditions.

(10) It is necessary that the manufacturing community have to look in detail at the formalized procedures of Group Technology, which can be systematically implemented to obtain greater benefits.

4. 翻译下面的句子，注意各种名词性从句的翻译。

(1) Precisely what happens to the atomic structure of metals at such low temperature is still not wholly understood.

(2) What the image tube does is not hard to describe.

(3) What you discovered will be a new rare element.

(4) That heat flows from a hotter to a cooler body is a process of energy transfer tending to equalize temperature.

(5) Whether life can exist on any planet other than the earth is very doubtful.

(6) The second is to endeavor to ensure that in the event of an accident there is the maximum possibility of survival of the occupants.

(7) The project on emerging Nanotechnologies estimates that as of August 21, 2008, over 800 manufacturer-identified nanotech products are publicly available, with new ones hitting the market at a pace of 3—4 per week.

(8) My aim is to describe what natural history survey was in its heyday, the reasons it flourished where it did, and how it worked in practice.

(9) Pure science consider how life has developed into the plants and animals we see on earth today.

(10) Plastics are different from other materials in that they possess a combination of properties.

(11) Another limitation of this history is that it treats mainly vertebrate zoology and some botany, but insects and other invertebrates hardly at all.

(12) That is why there is no life on moon.

(13) One of the remarkable things about it is that the electromagnetic waves can move through great distances.

(14) Perhaps the most common classification of material is whether the material is metallic or non-metallic.

(15) One of the significant fringe benefits of the remarkably small size of integrated systems is that effective external nuclear radiation shields are now feasible.

(16) There is a possibility that the thickness of filler plates for front and rear mount is different.

(17) The fact that electric currents passing through a wire produce heat is known to all.

(18) The suspicion that smoking has something to do with cancer of lung rests on several kinds of evidence.

(19) There is no certainty that the elastic material does not necessarily obey Hooke's law.

(20) There is every indication that algae, along with bacteria and certain fungi, are extremely ancient organisms.

5. 翻译下面的句子，注意定语从句的翻译。

(1) The other scale in general use nowadays is the binary, or two-scale, in which numbers are expressed by combinations of only two digits, 0 and 1.

(2) The most hazardous phases of a light are those in which the aircraft is close to the ground: takeoff, initial climb, final approach, and landing.

(3) Thinking of lightning, all the electrical energy in it is gone in a flash—changed into brilliant light which you can see, into heat which burns whatever is struck by the lightning and into sound which you can hear as thunder.

(4) The application of biotechnology is going to produce a set of possibilities that we simply cannot conceive of, even in our most imaginative flights of fancy.

(5) For farmers, crops engineered with genes that resist cold, drought, or other adverse weather conditions can boost crop yields with less money and effort.

(6) Quantum mechanical effects are very important at this scale, which is in the quantum realm.

(7) Further applications which require actual manipulation or arrangement of nanoscale components await further research.

(8) The 500-hPa height field shows a blocking high over the Gulf of Alaska, which lasts another seven days.

(9) Biodiversity is a term that describes the number of different species that live within a particular ecosystem.

(10) Then there is electricity, which is yet another sort of energy.

(11) There are some things about energy that are difficult to understand.

(12) A system is a collection of hardware, software, data, and procedural components that work together to accomplish an objective.

(13) Photographs are taken of stars, the light of which is too faint to be seen by eyes at all.

(14) Some materials, such as cotton, which is often used as insulation, are liable to absorb moisture, and this will adversely affect their insulating properties.

(15) Scientists are able to draw from these germs a substance which is a germ destroyer.

(16) Over-production of any commodity can create difficulties, because it can lead to a glut on the market, which may cause prices to fall sharply.

(17) Birds and animals which hunt at night have eyes which contain few or no cones at all, so they cannot see colors.

(18) As is shown in the table, the U. S. A is the major copper-producing country.

(19) The center of gravity of a passenger ship should be as low as possible, the initial stability of which may be greatly increased.

(20) One day on the moon is as long as two weeks on the earth, which has been proved by astronomers.

6. 翻译下面的句子，注意状语从句的翻译。

(1) Nevertheless, when a serious accident does occur it is usually catastrophic and receives widespread publicity, often out of proportion to the actual damage.

(2) When you think you know what energy is, suddenly it has changed into a totally different form.

(3) As they open and close cell membranes or travel through tissue and enter cells and viruses, machines will only be able to correct a single molecular disorder like DNA damage or enzyme deficiency.

(4) The sun produces nuclear energy from hydrogen gas and, day by day, its mass gets less, as matter is converted to energy.

(5) Where water resources are plentiful, hydroelectric power stations are being built in large numbers.

(6) The materials are excellent for use where the requirement of the workpieces is not high.

(7) Nanodevices could be observed at work inside the body using MRI especially if their components were manufactured using mostly 13C atoms rather than the natural 12C isotope of carbon, since 13C has a nonzero nuclear magnetic moment.

(8) With molecular machines, access to cells is possible because biologists can insert needles into cells without killing them.

(9) The doctor will also be able to scan a section of the body, and actually see the nanodevices congregated neatly around their targets (a tumor mass, etc.) so that he or she can be sure that the procedure was successful.

(10) The temperature of the explosion in the cylinders is so high that the metal of the cylinder walls cannot stand it.

(11) If the water is heated, the particles move more quickly, which agrees with our theory.

(12) This is because it is easier to perform the various mathematical operations if decimals are used instead of fractions.

(13) Just as 26 letters combine to form the words of our language, 20 amino acids combine to

form all proteins found in nature.

(14) In order that a rocket may put a satellite in orbit, it must attain a speed of about five miles per second.

(15) Many species are becoming extinct through deforestation, pollution, and human settlement, although others remain to be discovered.

(16) Though technologies branded with the term "nano" are sometimes little related to and fall far short of the most ambitious and transformative technological goals of the sort in molecular manufacturing proposals, the term still connotes such ideas.

(17) While it can't yet be said that every mouthful of food has been changed through genetic engineering, it is likely that almost every American has had a mouthful of engineered food.

(18) And though few Americans sit down to a plate of soybeans for dinner, these legumes arrive through the back door as an additive to scores of foods like mayonnaise, margarine, cooking oils, salad dressings, coffee creamers, beer, cereals, candy, and shortenings.

(19) The higher the internal pressure, the greater the tendency for the gas to expand.

(20) The more pristine a diverse habitat, the better chance it has to survive a change or threat—either natural or human—because that change can be balanced by an adjustment elsewhere in the community.

7. 翻译下面的句子，注意各种否定句的翻译。

(1) There are few to support the view that in a highly efficient transit operation almost every vehicle should be capable of running almost every service.

(2) Aside from the fact that electrons are too small to be seen, we could find it impossible to count them as they flowed by.

(3) The motor didn't stop because the electricity was off.

(4) At no time and under no circumstance will China be the first to use nuclear weapon.

(5) There is no steel not containing carbon.

(6) If Einstein didn't have all these qualities, he could never have done what he has done or have the chance of succeeding in what he is trying to do.

(7) In the absence of electricity, large scale production is impossible.

(8) The vibrations are not often visible but they occur none the less.

(9) Both of the instruments are not precision ones.

(10) The inactive gas neither burns nor supports combustion.

8. 翻译下面的复杂长句，注意不同翻译方法的应用。

(1) The main operations are: to add, subtract, multiply and divide; to square, cube or raise to any other power; to take a square, cube or any other root and to find a ratio or proportion between pairs of numbers or a series of numbers.

(2) Natural history survey arose out of a particular set of environmental, cultural, and scientific circumstances; ran its course; then gave way to new and different ways of studying natures diversity.

(3) Moreover, even an elementary knowledge of this branch of mathematics is sufficient to

enable the journalist to avoid misleading his readers, or the ordinary citizen to detect the attempts which are constantly made to deceive him.

(4) Scientists have variously estimated that there are from 3 to 30 million extant species, of which 2.5 million have been classified, including 900,000 insects, 41,000 vertebrates, and 250,000 plants; the remainder are invertebrates, fungi, algae, and microorganisms.

(5) The need for tunnel calibration that documents pressure coefficient, velocity vector magnitude and direction, acoustic disturbances that is discussed in this report lends support for further work on determining data accuracy and flow quality requirements for wind tunnels.

(6) This new capacity will allow for a wide variety of inexpensive and ordinary space experimentation, opening new doors for those who have been up to now "imaginative student astronauts".

(7) Nanotechnology is very diverse, ranging from extensions of conventional device physics to completely new approach based upon molecular self-assembly, from developing new materials with dimensions on the nanoscale to investigating whether we can directly control matter on the atomic scale.

(8) Most applications are limited to the use of "first generation" passive nanomaterials, including titanium dioxide in sunscreen, cosmetics, surface coatings and some food products; carbon allotropes used to produce gecko tape; silver in food packaging, clothing, disinfectants and household appliances; zinc oxide in sunscreens and cosmetics, surface coatings, paints and outdoor furniture varnishes; and cerium oxide as a fuel catalyst.

(9) All hydraulic systems and components which are subjected, during the operation of the aircraft, to structural or other loads which are nor of hydraulic origin, shall withstand such loads when applied simultaneously with appropriate proof pressures as specified in Table 1, without exceeding the yield point or normal load endurance limit of material at the maximum operating temperature.

(10) Accidents may be caused by: malfunctioning of the aircraft itself, such as structural failure of the powerplant or airframe, or incorrect system functioning; operational difficulties, such as adverse weather or air traffic conflicts; terrorism, where the major preventive effort has to be the initial checking of occupants and baggage; "acts of God", such as bird and lightning strikes or crew in disposition, in which case the aim is to minimize the consequences of the events.

三、拓展阅读和翻译练习

阅读以下科技语篇，并翻译划线句子以及黑体段落。

Petroleum

1. Petroleum is the largest source of liquid fuel, and, in spite of attempts to develop synthetic fuels, and the continued use of solid fuels, world consumption of petroleum products is about four times greater now than in 1940.

Crude petroleum oil from different oilfields is never exactly identical in composition. 2. Although all petroleum is composed essentially of a number of hydrocarbons, they are present in varying proportions in each deposit, and the properties of each deposit have to be evaluated.

Samples are subjected to a series of tests in the laboratory, the object of which is largely to determine the correct processing methods to be adopted in each case.

Petroleum is not normally used today in the crude state. 3. The mixture of oils of which it is composed must be separated out into a number of products such as petrol, aviation spirit, kerosene, diesel oils and lubricants, all of which have special purpose. The main method of separation used in refineries is fractional distillation, although further processing is normally required to produce marketable petroleum products. The different hydrocarbons present in petroleum have different boiling temperatures, and the fractions can therefore be isolated according to their boiling temperatures. Petrol, for instance is a mixture of the lower-boiling hydrocarbons, with boiling temperatures ranging from 100 ℃ to 400 ℃. Diesel oil on the other hand have boiling temperatures of upwards of 400 ℃.

Distillation was originally carried out in batch-stills and, although this is still done for special purposes, the development of the pipe-still has revolutionized refinery processes, since it allows continuous vaporization and rectification of the fractions. The pipe-still consists of a brick-lined furnace, in which is fitted a battery of tubes, through which the crude oil is pumped. The oil is heated, and petrol vaporization occurs. The oil then enters the fractionating tower, where it is distilled by coming into contact condensed vapour which has previously been evolved from the still. Fractions of different boiling ranges are drawn off at different points in the tower, or, in some plants, in a series of towers, each one distilling successively heavier fractions.

The heavier distillates, such as gas oil, undergo various other processes, of which the most important is known as cracking. 4. In this process, they are heated to a temperature of about 550 ℃, as a result of which the heavier molecules are broken up, lighter oils such as petrol being produced. 5. Catalytic cracking, in which silicon compounds are used as catalysts to aid the process of decomposition, gives higher octane petrol. These are widely used as motorcar fuels, since the high-octane value reduces the tendency of the fuel detonation.

Source: A coursebook for science and technology translation

第7章 语篇的翻译

在前面的章节我们详细讨论了科技英语词语和句子的翻译方法和技巧。从本章开始，我们将从语篇的层面探讨科技英语翻译。

本章节包括两部分的内容：一是介绍中文科技语篇的文体特点；二是讲解英文科技语篇翻译的基本步骤和翻译要点。

第1节 中文科技语篇的文体特点

根据翻译的对等原则，一篇英文科技语篇翻译过来自然应该是与其对等的中文科技语篇。因此，要做好语篇翻译，首先有必要了解中文科技语篇的文体特点。

中文的科技文体又称科技汉语，是汉语在科学技术领域内使用的一种文体。它的语言材料是现代汉语，相比其他文体，科技汉语又有自己的特点。同英语科技语篇一样，汉语科技语篇也同样具有逻辑严密、表达客观、行文准确精练、重点突出这样的文体风格。下文将从词汇与句法方面论述科技汉语的文体特点。

1.1 科技汉语的词汇特点

（1）大量使用科技术语。

科技汉语中所用的科学术语都具有精确的含义，每个术语在某一领域内只有一个严格规定的意义，因而具有单义性。在科技领域内通常不使用日常生活中的普通词语，因为普通词语有时含有几个意义，在表达中容易引起歧义。但科技汉语中同样有一些普通词语应用于科技领域中，也具有了术语的意义，例如：“质量”这个词一般表示“产品或工作的优劣程度”。在物理学中，“质量”是物体所具有的一种物理属性，是物质的量的量度，是一个正的标量。又如，“运动”这个词一般表示体育活动，或政治、文化、生产等方面有组织、有目的的群众性活动。但在物理学中，“运动”指一个物体和其他物体之间相对位置的变化。

（2）不带感情色彩。

科学技术反映的是客观事物、现象及其内在规律，是在生产和科学实验的基础上获得的，是不以人的意志为转移的。因此，科学术语不能，也不应该带有喜怒哀乐的感情色彩。

（3）绝大多数科学术语是名词或名词性词组。

据统计，《辞海》（理科分册）“物理”类共收术语 1 670 条，其中 1 500 多条是名词或名词性词组，其他 100 多条是量词、动词和形容词。例如：

- 场 功 始 动 能 熔点 宇称 压 强 阻 力 冲量（名词）

- 质点组　参照　系　交流　电　电场强度　万有引力　瞬时速度　自由落体运动（名词性词组）
- 埃卡　焦耳　库仑　高斯　达因　（量词）
- 磁化　跃迁　传导　并联　辐射　（动词）
- 宏观　微观　自动　绝对　无穷　（形容词）

另外，在科学术语中，双音节的名词占多数，这与现代汉语词汇发展的总体趋势是一致的。

(4) 大量使用抽象词。

抽象词指的是具有高度概括的理性意义或语法意义，而没有感情色彩和形象色彩的词语。科技文体中大量运用这些抽象词，按照现代汉语的句法结构遣词造句，将科学术语串联起来，说明自然现象，阐明科学道理，表达逻辑思维，而成为有丰富内容的科学语篇。而与此相反，文艺作品则广泛使用具有感情色彩和形象色彩的词语。例如：

……的剪纸构思新颖，刀法巧妙，人物栩栩如生，充满浓厚的生活气息，给人以美的享受。

我从彼得手里接过这件珍贵的礼物，非常感动。我珍惜这件礼物，更珍惜彼得对中国人民的友谊。

“运载火箭发射成功了！”人们欢呼着、目送着腾空而去的火龙，心潮像海浪一样翻滚。

由此可见，科技语篇和文艺作品中使用的普通词语，其情况不尽相同。

(5) 汉语科技文体一般不使用下列词语：

- 叹词、象声词
- 歇后语、谚语和方言土语，成语也用得较少
- 一部分语气助词，例如：啊、吧、么、呢等
- 一部分副词，如：明明、大肆、断然、极力、万万、分外等
- 一部分儿化的词，如：花儿、鸟儿、劲儿等
- 形容词的重叠形式，如：好好儿、白茫茫、好端端等
- 动词的重叠形式，如：商量商量、讨论讨论等
- 描写人物性格、品质的词语，如：高尚、刚强、聪明、果断等
- 其他词，如：喜欢、号召、动员、捍卫、害怕等

这是因为科技文体是属于比较正式的文体，上述这些词语或者过于口语化，因而有损文体风格；或者表现过多的感情色彩而影响文体的客观性。

(6) 汉语科技文体中保留了一定数量的文言词语。

文言词语是古代文献中流传下来的，其中一部分至今仍在科技文献中广泛运用。文言词语主要用在科技文体的书面语中，以使其正式庄重，并使语言表达得精炼简洁。例如：

内能的改变量只决定于初末两个状态，而与所经历的过程无关。

(7) 大量使用虚词。

为了表达丰富的科技内容和复杂的逻辑关系，科技文体中多用长句和复句，而连词、介

词和具有关联作用的副词是组织长句和复句必不可少的手段，所以科技文体中常用这些虚词。例如：

气体分子的热运动是永不停息的。如果一定量气体在一定容器中，与外界没有能量的交换，内部也没有任何形式的能量转换（例如，没有化学变化或原子核反应等），则不论气体内各部分的原始温度和压强如何，经相当时间后，由于分子的热运动和相互碰撞，终将达到密度均匀、温度均匀和压强均匀的热动平衡状态。

1.2 句法特点

（1）汉语科技文体中，主语是客观现象、事物时，多用完全的主谓句；科技论文强调结构紧密、层次分明、论证严谨、说理清楚，要求遣词造句避免发生歧义、引起误解，因此多用完全主谓句。例如：

- 物质从固态变成液态的现象叫作熔解 。

但是，当主语泛指人时，往往可以省略。例如：

- 如果要测量部分电路中的电流强度，必须把安培表串联在这部分电路里。

这是因为自然科学叙述的是自然界中的普遍真理，不论是谁，都应该按客观规律办事，所以不必在句子中具体指明人称。

（2）多用陈述句。

汉语科技文体主要用来叙述自然界的客观规律及其应用，因此多用陈述句。又因为这种文体一般不用来抒发人的感情，所以几乎不用感叹句。疑问句多用于习题中，在正文中用得很少。祈使句常见于命题、解题、实验指导、操作说明等方面。从语序上看，基本上用正常语序，几乎不用倒装句。

（3）单句的复杂化。

在汉语科技文体中，单句比较复杂，主要是因为：定语和状语等附加成分的增多以及复合成分的运用。

定语和状语的使用

如上所述，科技文体中使用的多是抽象概括的词语。为了深刻揭示事物的特殊矛盾，区别事物之间的界限，准确地表达和判断，需要从数量、性状、质料、范围和所属等方面限定名词所表示的事物，从时间、空间、范围、情态、程度、条件和方式等方面来限定动词所表示的动作变化，限定形容词所表示的性质或状态，因此充当定语和状语的往往不是一个词，而是几个词，尤其是常常使用各种结构形式的词组，于是形成了科技汉语中的长定语和长状语。例如：

- 示波管的控制极是一个套在阴极前面的金属圆筒。
- 加在密闭容器内的液体上的压强，能按照原来的大小由液体向各个方向传递。

分析：

在前一个例句中，“筒”前边有表示数量、处所、质料、性状的定语。而在后一个例句中，“加在密闭容器内”和“液体上的”递加作“压强”的定语。三个介词结构：“按照原来的大小”“由液体”和“向各个方向”并列作谓语动词“传递”的状语。

词和各种结构形式的词组在句子的附加成分内，错综编插，层次繁复，或并列等立，或递加相随，或交错堆叠，这样就使定语和状语等附加成分不断地增多和延长，从而使单句的结构变得复杂起来。

而且在汉语科技文体中，使用的定语多是限制性的，是用来限制或规定它所说明的词语的意义范围的。而在文艺作品中，定语多是修饰性的、描绘性的。在汉语科技文体中，定语和状语在表达中的作用十分重要，它们可以使定义、定理、定律、假说和现象等表述得精确细密；不用或少用这些附加成分，往往会造成概念上的模糊和对理论的误解。

复合成分的运用

汉语科技文体表达的是逻辑思维，即从大量的现象材料中，概括归纳出本质的规律来。因此科技文体中常常使用并列结构作句子的各种成分，这就是复合成分。例如：

- 氮气的状态、颜色、气味、溶解性、沸点等，不需要发生化学反应就能表现出来。
- 牛顿第二定律确定了力、质量和加速度之间的关系。
- 简单电路的计算，通常是应用电阻定律、欧姆定律和导体串联或并联的知识来分析、计算电流强度、电阻或电压。

（4）大量使用复句。

客观事物不断地发展变化，其间的关系极其错综复杂。随着科学技术的进步，人类对自然现象及其规律的认识日益深入，用来表达思维逻辑的句法结构越来越精密化、复杂化。复句，特别是偏正复句是表达精密复杂的思维关系的，因此在汉语科技文体中大量运用。而且，为了描述客观事物间的更为错综复杂的关系，还使用多重复句。多重复句各分句间的层次关系比较复杂，既有联合关系又有偏正关系。

1.3 中英科技文体在句法上的差异

中英科技语篇虽然整体风格相似，但是在句法层面还是存在各自特点的。例如，从形式上看，汉语描述过程与发出指令时采用的句式差异不大，而英语则必须用陈述句描述过程，用祈使句来表达指令；英文科技语篇中大量使用非谓语动词、后置定语、定语从句，以及形式主语句等结构，但在汉语中则没有完全对等的形式。而在那些大致相同或者相似的地方（如被动句式、名词化、长句等），中英文科技文体也存在微观上或者频率及量级上的差异，这些差异对翻译的处理也有着不可忽视的影响。

首先，在被动的表达上，中文科技语篇的应用频率远低于英文。因此，许多英文的显性被动句式往往要翻译成中文的隐性被动句式，或者译成无主语句式。

在名词化方面，英文的名词化倾向远比中文突出，而动词优势却是中文的一大句法特点，对于科技语篇也不例外。由于名词化在英汉两种语言的科技文体中有不同的存在方式，两者的名词化形式也难以一一对应。因此，翻译过程中，译者往往需要将其降解为动词占优势的中文表达形式。

在长句方面，英文多为并列句或者复合句，呈树形结构，以主谓结构形成句子的主干，其余成分无论多复杂都依附于主干上；而中文的长句则有两种情况：一种是用句号、问号或者感叹号作为其结束标志的语言片段。其中不一定有统领全句的主谓结构，却可能有若干具备完整主谓结构的语言片段，即小句，它们之间以逗号间隔，一逗到底终成一句；另一种情况则可能以主谓划分，但不一定用句号、问号或者感叹号作为其终结的标志。无论哪种情况，其显著特点之一是前置定语的长度非同一般。

只有充分了解中文科技语篇的文体特点，在翻译英文科技语篇时，才能根据具体的待译文本、目标读者的语言背景来采取灵活的技巧方法进行转换，将其翻译成与原文对等的中文科技语篇。

第2节　英文科技语篇的翻译

在掌握了一定的科技英语词语和句子的翻译方法和技巧之后，是不是译好每一个词语和句子，再将它们拼接起来，就能够很好地完成一篇科技文章的翻译呢？答案当然是否定的。要想高质量地完成英文科技语篇的翻译，我们必须要学习和掌握一些科技语篇的翻译方法和技巧。

2.1　篇章翻译的步骤

进行英语科技篇章翻译，一般可以遵循以下几个步骤：

（1）通读全文后，确定文章学科类别。树立“大语篇观”的思想，即对该语篇的相关专业领域、背景知识、基本理论等有较好或者至少是粗浅的了解，为后期翻译工作打下良好的基础。

（2）以原文自然段为基本单位进行翻译。首先，精读该段落，把握段落中上下文之间的内在逻辑关联，深度理解原文。然后，逐句进行翻译，注意句中专业术语的正确处理。翻译过程中，有时可考虑句与句之间的逻辑关系，对句序进行适当调整。

（3）完成所有段落的翻译后，对篇章进行整体把握，注意考虑段落内及段落间的衔接和连贯。

（4）以“忠实、通顺、规范”作为衡量标准，通读核查译文，并进行修改和完善。注意用词和语气的前后一致性，并检查日期、时间及数据等重要细节。

2.2　篇章翻译要点

（1）标题的翻译。

标题揭示文章主题、概括文章内容，用词需简明扼要。我们通常将标题比作文章的“窗口”或“眼睛”，因此标题翻译的好坏对译作而言十分重要。科技英语文章标题的翻译要求严谨准确、紧凑醒目。

科技英语文章标题主要采用短语结构的形式，少数采用句子以及主副标题的形式。我们前面讲过的一些翻译方法在翻译标题时也是适用的，下面通过具体实例介绍不同类型标题的翻译方法。

短语结构的标题

短语结构的标题最为常见，其特点是短小精悍、中心词突出、意思一目了然，翻译成汉语后我们仍然需保留以上特点。

①名词短语结构。

前置定语 + 中心词

这种结构的标题很常见，通常可以译为汉语的偏正结构。例如：

The Universe's Invisible Hand

宇宙中的隐形手

A Solar Grand Plan

太阳能利用的长远规划

Terrorism's Changing Characteristics

恐怖主义不断变化的特征

名词或名词并列词组结构

这种由名词或名词成分的并列结构构成的标题也很常见，翻译起来并不困难，一般顺译即可。例如：

Knowledge Management

知识管理

Climate and Human History

气候与人类史

The Space Shuttle and Geological Remote Sensing

航天飞机与地质遥感

中心词 + 后置定语

这种结构的标题尤为常见，后置定语可以是介词短语、分词短语、定语从句、动词不定式或副词等多种形式。一般情况下翻译时可采用倒译法将后置定语逆序译成中心词的前置定语。例如：

An Engine for Tomorrow's Small Helicopter

用于未来轻型直升机的发动机

"The Light of Reason" or "the Light of Ambiguity"

"理性之光" 还是 "模糊之光"

An Analysis of Issues Concerning "Acid Rain"

"酸雨" 问题分析研究

Adaptive Distance Protection Based on R-L Model Error

基于 R-L 模型误差的自适应距离保护

Insulator That Can Self Clean
可自洁的绝缘子
Measures to Remove Dust
除尘措施
International Electricity Market Today
当今国际电力市场

②非谓语动词结构。

非谓语动词结构构成的标题主要以动名词结构和不定式结构最为常见。一般可采用顺译的方法翻译。例如：

Mapping Out Future of Intelligent Robots
规划智能机器人的未来
How to Reduce Stress
如何减压

③介词短语结构。

在英语科技文章中，有些文章的标题是“介词 + 中心词”结构，翻译时注意介词的正确译法，例如：

On the Drag Reduction of the Shark Skin
论鲨鱼皮革的减阻能力
Towards Times of Electric Packaging
走进电气化包装时代

句子作标题

有些标题由完整的句子构成，可以是陈述句也可以是疑问句，但一般句子都较为短小。翻译时可参考前面讲过的句子翻译的一般方法进行翻译。例如：

Study Suggests Primates and Dinosaurs Shared the Earth
研究表明灵长目动物与恐龙曾在地球上共存
Could the Internet Ever Be Destroyed?
因特网会有灭顶之灾吗？
What Happens During a Nuclear Meltdown?
核泄漏会发生什么事情？

主副标题

如果标题内容中包含不同层次可以用主副标题的形式来体现，即利用副标题对主标题进行解释或补充说明。英语中的副标题可以用冒号引出，也可以由破折号引出，翻译成汉语时可以根据情况灵活处理。例如：

Application of Nanotechnology in Perfumes: Thrills and Treats of Smelling Nano
纳米技术在香水制造中的应用——气味纳米带来的兴奋与威胁

（2）科技术语的翻译。

科技语篇的一大典型特征是大量使用专业术语，且文章涉及领域专业性越强，其中专业术语使用就越多。科技词汇和术语用在某一专业范围内词义一般比较稳定，但是也会出现有些词汇具有专业术语和日常普通用语的双重属性（半科技词汇）或者同一词汇在不同专业领域中的意义不同的情况（通用技术词）。在翻译时，我们一定要通过仔细的通篇阅读，把握文章的专业理念，找准对口专业领域，确定文章中涉及的专业词汇在特定的专业领域中具有的特定的意义，切勿将科技词语误译成不具有特殊专业含义的普通词语。而且，在翻译过程中应严格遵循某一专业领域的用语习惯，某一词语一经译出，即应保持一贯性，在上下文中不应随意改动，以免引起概念上的混乱。

例文：

A **memory** is a medium or device capable of storing one or more bits of information. In binary systems, a bit is stored as one of two possible states, representing either a 1 or a 0. A flip-flop is an example of a 1-**bit memory**, and a magnetic tape, along with the appropriate transport mechanism and read/write circuitry, represents the other extreme of a **large memory** with an over-billion-bit capacity.

Computer memory can be divided into two sections. The section common to all computers is the **main memory**. A second section, called the **file or secondary memory**, is often present to store large amounts of information if needed.

The main memory is composed of semiconductor devices and operates at much higher speeds than does the file memory. Typically, a word or set of data can be stored or retrieved in a fraction of a microsecond from the main memory.

We shall limit our discussion to semiconductor main memory. There are two broad classifications within semiconductor memories, the **read-only memory (ROM)** and the **read-write memory (RWM)**. The latter is also called a RAM to indicate that this is a **random-access memory**.

分析：

通过阅读我们可以发现“memory”一词是这篇文章的主题词，该词出现频率非常高。正确译出该词是把握全篇核心内容的关键所在。“memory”一词属于半科技词汇，在日常用语中的意思是“记忆”，但在这里绝不能照搬此意，而应根据篇章所涉及的相关学科领域，译为专业术语“存储器”。另外，篇章中多次出现和“memory”相关的词语，如：“1-bit memory, large memory, main memory, file memory or secondary memory, read-only memory (ROM), read-write memory (RWM), random-access memory”等。在翻译时，我们要坚持一贯性原则，将上面和“memory”相关的概念保持一致地译为“一位存储器、大存储器、主存储器、文件存储器或辅助存储器、只读存储器、读写存储器和随机存取存储器”，从而保证了译文的科学性和专业性。

译文：

存储器是能够存储一位或多位信息的媒体或装置。在二进制系统中，一位以两种可能状态之一进行存储，分别代1或0。触发器就是一位存储器的例子。配有合适的传送装置和读写电路的磁带是大存储器的另一个极端的例子，存储能力在10亿位以上。

计算机的存储器可以分成两部分。所有计算机都有的部分是主存储器。第二部分称为文件存储器或辅助存储器，在需要的时候，常用以存储大量的信息。

主存储器是由半导体器件组成的，其运行速度比文件存储器快得多。一般来说，以零点几微秒的时间即可对主存储器存或取一个字或一组数据。

我们只讨论半导体型主存储器。半导体型存储器分为两大类：只读存储器（ROM）和读写存储器（RWM），后者也称为RAM——随机存取存储器。

（3）语篇的衔接（Cohesion）和连贯（Coherence）。

语篇是实际使用的语言单位，是一系列连续的话段或句子所构成的语言整体。构成语篇的词语、句子和段落等成分都不是孤立存在的，它们在形式上是衔接的，在语义上是连贯的。由此可见，衔接和连贯是语篇的重要特征，是语句形成段落、段落构成语篇的重要保证。

翻译一篇文章就好比建造一座大楼。在用一片片砖瓦开始建造房屋之前，必须先做好一个整体的设计和规划。翻译实践活动也要从词和句子开始，但要做好翻译，译者必须以文本的整体概念和思想为指导，把握文章的全局脉络，深度分析和理解上下文之间的逻辑关联，对原文篇章中出现的衔接和连贯等叙述策略予以精准的、汉语化的处理。这样才能将原文转化为符合汉语思维逻辑的、能够准确传达原文内涵信息的、语句流畅通顺的一篇汉语译文。

语篇的衔接（Cohesion）

衔接是指使语篇得以存在的语言成分之间的关联关系。衔接主要体现在语篇的表层结构上，它通过语法和词汇等看得见的衔接手段使行文流畅完整、文脉相通。可以说衔接是语篇的有形网络。

①语法衔接（Grammatical Cohesion）。

语法衔接手段包括照应、替代、省略和连接，下面分别举例说明。

照应（Reference）

照应是英文科技语篇衔接手段中广泛使用的一种衔接手段，它是指用简短的指代来表达上下文中已经或即将提到的内容，使用照应可使语篇在结构上更加紧凑，成为前后衔接的整体。照应包括人称照应（通过人称代词 he，it，they，them 等来实现）；指示照应（通过指示代词 this，that，these，those 和定冠词 the 来实现）和比较照应（通过表示比较事物异同的形容词或副词和其比较级来实现，如 same，equal，identical，such，different，another，other，better，more，less 等）。在翻译过程中，由于英汉语言存在不同的思维方式和表达形式，遇到上下文出现照应的关系时，应灵活处理。例如：

Microcomputers—also called personal computers—are small, self-contained computers that fit on a desk-top and are usually only used by one person. **They** are the least expensive type, and are widely used in businesses for a variety of tasks, such as word-processing, small data-base management, and spreadsheets. **They** are also used as home computers, for family budgeting and similar jobs, as well as for games.

分析：

该段落中多处用人称代词进行照应。由于第一句中有一个破折号引出的插入语部分，句子较长，翻译时采取分译的方式译为两句话，第二句增译主语“它”，指代“微型计算机”。这里注意，虽然原句用的是复数概念，但因为复数在这里表示泛指一类事物，所以不需要用“它们”一词来译。后面两句中的主语均用“they”来指代前面第一句中的主题词“microcomputers”。翻译时，如果这两句都用“它”一词来翻译，则显得指代不清，不符合汉语表达。因此，在这里要灵活处理，分别用“这类计算机”和“此类计算机”来翻译。

译文：

微型计算机也叫个人计算机。它是一个小型的、独立的安装在桌子上的计算机，通常只由一个人使用。这类计算机是最便宜的一种，广泛应用在商业领域，用于完成各种各样的任务，比如文字处理、小型数据库管理和电子表格。此类计算机也供家庭使用，用于做家庭预算和类似的工作，也可用来玩游戏。

The author has selected **those products** that demonstrate basic principles and also **those** that he has used personally with success in plant or laboratory services either in the United States or in Germany.

分析：

原文中用指示代词“those”照应前面的“those products”，翻译时要译为“那些……的产品”。这样的照应手段不仅使上下文浑然一体，也有效避免了重复。

译文：

这位作者选择了那些能说明基本原理的产品，还选择了那些他本人用过的产品，这些产品曾在美国或德国的设备或实验室中成功地使用过。

James Hopwood, British mathematician noted for his investigations into the relationships between mathematical concepts and the natural world. **Equally**, Peano enjoyed popularity in the mathematical area for his induction axiom.

分析：

原文中用副词“equally”一词与前句形成比较照应，将两句紧密连接在一起。翻译时，为了保持信息的对等，用汉语“同样”一词翻译，保留了照应的衔接手法。此外，汉语中常用“同样的、同等的、类似的、不同的”等形容词以及“如……一样、和……差不多、不像……那样”等比较结构来实现前后的比较照应关系。在这方面，英汉两种语言的用法差异不大，翻译起来不太困难。

译文：

詹姆斯·霍普伍德，英国数学家，以探索数学与自然界之间的关系著称。同样，皮诺亚也因归纳数理在数学界享有盛誉。

替代（Substitution）

替代是指用一种表达来替代另一种表达的语句间的衔接手段。使用替代可以使科技语篇行文连贯，避免重复。常用做替代词的词语有 one，same，do，so 等。例如：

In the past decades, many experiments proved that this method cannot help increase the output of the steel. But **the one** conducted by the Steel and Iron Center indicated that the method can double the yields.

分析：

原句中的 one 一词替代上句中的 experiment 这一概念，但不替代 experiment 前面的修饰成分，也不包含复数的概念。翻译时只需要译成“实验”，体现被替代的概念本身即可。

译文：

在过去的几十年里，许多实验都证明该方法对提高钢产量无效。但是，钢铁中心的实验却表明这种方法可使钢产量增加一倍。

省略（Ellipsis）

英语语篇中经常出现省略的现象。通过省略某些上文中提到或出现过的词来达到上下文衔接的目的。省略的部分可以是句子的某个成分，也可以是整个小句。在翻译时，常常需要将原文中省略的词语补充出来，使译文表达更加准确、流畅，符合汉语表达习惯。例如：

Some diseases are communicated from man to man only; some, from animal to man; some, from man to animal and from animal back to man; and some, from animal to animal only.

分析：

很明显，英语原文中多次省略了名词“diseases”和动词结构“be communicated”。原文中的省略不但不影响意思的表达，反而增强了语句之间的衔接。但是，如果按照英文的省略的表述方式直译成汉语，就会造成语义不清、理解困难的问题。因此，我们要把省略的词复现出来，以符合汉语表达习惯。

译文：

有些病只在人与人之间传播；有些病从动物传到人；有些病则从人传到动物，又从动物传回到人；而有些病则仅在动物之间传播。

连接（Conjunction）

运用逻辑连接词连接语篇是英文科技语篇中重要的衔接手段。逻辑连接词体现句与句之间的逻辑思维，使语篇结构更加严谨。科技篇章中常用的连接方式有四种：增补关系（and, also, or, in addition, besides 等）；转折关系（yet, but, though, however, on the other hand 等）；因果关系（because, so, therefore, consequently 等）和时间关系（and, next, first, finally, meanwhile 等）。翻译这些连接关系时，一定要先弄清楚上下文之间的逻辑意义，再根据汉语的行文习惯，采用保留、省略或转译等不同的翻译方法进行处理。例如：

Access to cells is possible **because** biologists can insert needles into cells without killing them. **Thus**, molecular machines are capable of entering the cell. **Also**, all specific biochemical interactions show that molecular systems can recognize other molecules by touch, build or rebuild every molecule in a cell, and can disassemble damaged molecules. **Finally**, the self-replication of cells proves that molecular systems can assemble every system found in a cell. **Therefore, since** nature has demonstrated the basic operations needed to perform molecular-level cell repair, in the future, nanomachine based systems will be built that are able to enter cells, sense differences from healthy ones and make modifications to the structure.

分析：

原文中用了大量的、不同类型的逻辑关联词将语句紧密地编织在一起，逻辑概念环环相扣，步步深入，高度体现了科技英语中推理论证的严谨性。翻译时，我们一定要首先辨认清楚各种概念间的确切含义，结合原文中使用的关联词，运用判断与推理的方法，理顺各种概念间的逻辑关系，真正掌握原文的思想内涵，这一点是做好翻译至关重要的一步。对于关联词的翻译，这里大部分可采取直译的方式。但要注意，在第一句的翻译中，可将“原因关联”转译成“结果关联”，使行文更为自然。与英语不同，汉语中“表因”和“表果”的关联词既可以单用又可以连用，翻译时可根据具体情况适当选择使用。另外，第二句的翻译，也可相应添加“而”一词，以增加与前句的关联度。

译文：

生物学家能够在不杀死细胞的前提下将针扎入细胞，从而解决了进入细胞的问题。而分子机器能够进入细胞的原理也正是如此。此外，所有特定的生化互动表明，分子系统通过分子间相互触碰能够识别其他分子，可以在细胞中建立或重建每一个分子，还能将损伤的分子予以移除。最后，分子系统能够对细胞中的所有系统进行组装，这一点通过细胞自我复制也得到了印证。所以，由于自然界已经证实了进行分子级细胞修复所需要基本操作的可行性，在未来人们将成功研制出基于纳米机器的分子系统，该系统能够进入细胞，检测出与健康细胞的差异，并对其结构进行修正。

②词汇衔接（Lexical Cohesion）。

通过丰富的词汇关系衔接语篇也是科技语篇中非常重要的特点。词汇衔接关系可分为两大类：复现（reiteration）和同现（collocation）。

复现（Reiteration）

词汇复现是指通过某个词的原词、同（近）义词、上下义词等在语篇中的重复出现，达到衔接语篇的目的的衔接手段。例如：

A **bipolar transmitter** consists of two P-N junctions in close proximity. For an **N-P-N transmitter**, the emitter-base junction is forward-biased and electrons are injected into the base region, producing an excess of minority **carriers** there. These **carriers** diffuse to the **collector** junction where they are accelerated into the **collector** by the field in the depletion region of the reverse-biased junction.

分析：

原文中出现了三组复现词汇，第一组 A bipolar transmitter 和 P-N transmitter 是同义词复现；第二组 carriers 和第三组 collector 都是原词复现。多次使用的复现手段将语篇上下文信息紧密地连接起来，逻辑关系清晰。翻译时，可采取保留原文复现关系的方式，将几组词汇准确地进行对等翻译。

译文：

两极发射机由两个相邻的 P-N 结组成。对于 N-P-N 发射机，其发射体基极结是正向偏置的，电子被发射到基极后，产生了多余的少数载流子。这些载流子向集电极扩散，由于受反向偏置的集电极基极结阻挡层电场的作用而加速向集电极运动。

Dynamics is one of the essential foundations of physics. The **subject** was first soundly established in the 17th century on the basis of axiom usually known as Newton's law of motion.

分析：

原文中 dynamics 是 subject 的下义词。使用上下义词汇复现的衔接方式，不仅可以使语篇组织更加严密，还可以使阐述的内容更加具体化。在翻译时，我们要保留原文中的这种词语上下义的关系，将两个词分别译为“动力学”和“学科”。

译文：

动力学是物理学研究的重要基础之一。这门学科最初是在十七世纪，以著名的牛顿运动定律为基础发展建立起来的。

同现（Collocation）

词汇同现又叫词汇搭配，它是指彼此之间有某种关系或联系的不同词汇在语篇中同时出现，或习惯性共现。同现的词汇包括互补词、反义词，可以出现在同一个句子中，也可以在不同的句子中。这种搭配出现能把分布在不同小句中的若干成分从语义上联系起来，起到衔接篇章的作用。例如：

Conventional wisdom has long held that a **machine's** abilities are limited by the imagination of its **creators**. But over the past few years, the **pioneers** of **EHW** (evolvable hardware) have succeeded in building **devices** that can tune themselves autonomously to perform better.

分析：

原文篇章中有几组构成搭配关系的词语，包括 machine 和 creators; creators 和 pioneers; EHW 和 devices。这些具有搭配关系的词语的同现将语义紧密连接，环环相扣。在汉译过程中，只有发现了这些词汇的相互关联，才能帮助我们找准词汇的定位并将它们正确地译出。比如：这篇文章中，要和 machine“机器”一词搭配，creators 则应该从它很多的词义，如“发明者、创造者、设计者”当中，选择“研发者”这一译文；creators 和 pioneers 词义具有相关性，因此应将 pioneers 译为“开发者们”而不是“先驱者们”；device 一词的对应词义很多，如“装置、设备、仪器、器械”等，但在这里要和前面 EHW 一词对应，应该译为“仪器”一词最为恰当。

译文：

长期以来，人们普遍认为研发者的想象力制约着机器的能力。但是在过去几年里，EHW(演化硬件）的开发者们已经设计出能够自动调试以执行更好操作的仪器。

语篇的连贯（Coherence）

连贯是指语篇中语义的关联，它存在于语篇的深层，通过上下文逻辑推理达到语义的连接。可以说，连贯是语篇的无形网络。在科技英语语篇中，“连贯”这一特征尤为突出。我们发现科技语篇常常有一个内在的的逻辑思维从头到尾贯穿全篇，将所有概念有机地串联在一起。因此，翻译时我们一定要注意把握语篇中的“连贯”，理清似乎相互独立，实为相互关联、相互照应的句内、句间和段落间的关系，然后按照汉语的思维逻辑和表达习惯组织语言，将原作的主旨和内涵充分表达出来。由于英汉两种语言逻辑思维和表达方式上的差异，有时候我们不能完全对原语进行等值翻译，这就需要我们按照正确的逻辑思维，调整语段内语句的表达顺序，重新构建和组织语句，从而形成一篇语义明确清晰、逻辑结构严密、语言表达流畅、符合汉语行文模式的译文。例如：

According to a growing body of evidence, the chemical that make up plastics may migrate out of the material and into foods and fluids, ending up in your body. Once there they could make you very sick indeed. That's what a group of environmental watchdogs have been saying, and the medical community is starting to listen.

原译：

越来越多的证据表明，许多塑料制品的化学成分可能会从塑料中释放出来，进入食物或流体之中去，最终进入人体。一旦进入人体，便会使人生病。这是一些环保检查人员所指出的问题，并且这一问题也开始受到医学界的关注。

分析：

上面的译文完全按照英语原文的表述顺序进行翻译，结果出现了语义不流畅、不连贯，缺乏逻辑关联性的问题。汉语的思维方式一般是顺向思维，即某人做了什么或发现了什么，继而说明结果怎样。因此，我们要对原文的语序进行适当调整，对文章结构进行重新组织安排。

改译：

一些环保检查人员指出，越来越多的证据表明，许多塑料制品的化学成分可能会从塑料中释放出来，进入食物或流体之中并最终进入人体。这些化学物质一旦进入人体，便会使人生病。现在，这一问题也开始受到医学界的关注。

Modern man pours a lot of nitrates into the rivers, lakes and seas. These come from the fertilizers used for crops and from the human wastes of the world's growing population. In a lake, for example, these nitrates cause a great growth in plant life. Soon there are too many plants and they start to die. Bacteria eat the dead plants. Then soon there are too many bacteria. They use up the oxygen in the water. In a time, all the oxygen in the water is gone. The bacteria, plants and fish all die. The lake is now dead; no life will grow in it.

原译：

现代人把大量的硝酸盐倒入河流、湖泊和大海中。这些来自用于庄稼的肥料和世界正在增长的人口的人类垃圾。在湖中，例如，这些硝酸盐使植物大量生长。不久，就有太多的植物，而且它们开始死去。细菌如果吃掉这些死了的植物，随后就会有太多的细菌。它们用完了水中的氧。于是水中所有的氧都没了。细菌、植物和鱼全死了。湖“死”了，没有生命会生活在那里。

分析：

上面这篇英语科普文章中的词汇和句子简单，意思浅显易懂，感觉没有太大的翻译难度。但是，读完上面的译文，我们发现尽管译者非常忠实原文地进行了翻译，没有任何语言上的误译，但译文却令人费解，前后语句逻辑不清。这是因为译文虽然在词、句上做到了忠实，却缺乏对原文内在逻辑关联性和连贯性（如上下文之间的因果关系）的呈现。由于英汉两种语言的差异，我们做翻译特别是篇章翻译时，一定要注意上下文之间逻辑关联的体现，必要时要按照汉语表达习惯，通过适当调整或重新安排句子顺序、增加逻辑关联词体现隐含的逻辑关联等方法，实现中文篇章的连贯。

改译：

在种植庄稼所用的化肥以及日益增长的世界人口所排放的垃圾中，有着大量的硝酸盐类化学物质。这些化学物质被排放到河流、湖泊和海洋之中。就拿湖泊来说，大量的硝酸盐类物质促使湖中植物快速生长，不久植物便因数量过多而开始死亡。这些死掉的植物继而被细菌分解，致使细菌数量急剧增加。过多的细菌又耗尽了水中的氧气，最终导致湖中的细菌、植物和鱼类全部死亡。这样，湖就成了“死”湖，没有任何生物能够在湖中生存。

在科技英语翻译中，由于英汉两种语言在词法、句法、衔接方式等方面存在差异，逻辑思维方式和行文方式也不尽相同，造成源语和目的语之间不能完全对等地进行翻译。这就要求译者在翻译过程中，在遵照原文的意图、还原原文内涵的基础上，做出相应的变通，按照正确的逻辑关系，对译文进行构建。翻译工作者在翻译实践活动中必须树立较强的宏观篇章意识，站在把控整个语篇的高度，以一种动态的视角来审视和组织译文，力求使译文意思明确清晰、文理通顺、语义完整、语言自然流畅、符合汉语表达习惯。

第3节　练习

一、思考题

1. 科技汉语在词汇层面上具有什么特点？
2. 科技汉语中为何会大量使用虚词？
3. 科技汉语中单句为何会呈现出复杂化趋势？
4. 科技汉语中为何要大量使用复句？
5. 英汉语言的语域变体对翻译可能的影响是什么？
6. 科技英文篇章翻译一般包括几个步骤？进行篇章翻译时有哪些环节应该特别注意？
7. 如何理解语篇的衔接和连贯两者之间的关系？

二、翻译实践

1. 翻译下面的标题。

(1) Study Finds Smog Raise Death Rate

(2) Key Military—Technical trends

(3) Electric and Other Lasers Advance in Development

(4) The Future of Mars: Plan for NASA's Next Decade of Red Planet Probing

(5) Nanoparticles + light = Dead Tumor Cells

(6) The Caste System in India

(7) Finding Relevant Information on the Internet

(8) Bush Sees Long Future for Autonomous Weapons

(9) Current State of First Generation strategic Information Warfare

(10) United Nations Convention on Contracts for the International Sale of Goods

2. 翻译下面的语篇，注意用词的一贯性。

Clouds can greatly affect the temperature of the earth's surface. When there are many clouds in the sky, all of the sun's rays cannot reach the earth. The cloudy day then, will be cooler than the cloudless day. Clouds also prevent the earth from cooling off rapidly at night. For this reason, countries such as the British Isles, which are often covered by clouds, have a relatively constant temperature. The weather in these cloudy areas is neither very hot in summer nor very cold in winter. On the other hand, places such as deserts, which have few or no clouds, have very sharp variations in temperature—between night and day as well as between summer and winter.

3. 翻译下面的语篇，注意语篇的衔接与连贯。

(1)

In the 100 elements available to us, about three quarters can be classified as metals. And, about half of these are of at least some industrial or commercial importance. Metals, be they pure or alloy, can be further defined as being ferrous or nonferrous in make-up. Ferrous alloys are those in which the base or primary metals is iron, manganese, chromium. All other metallic materials automatically fall into the nonferrous category.

(2)

Over the past sixty years the blast furnace process has made tremendous progress and attained remarkable performances. Nevertheless, the blast furnace is today strongly challenged by some new processes which should respond to the objectives of producing metallic iron directly from ore fines and from non-coking coals without polluting the environment. These new processes are mainly direct reduction and smelting reduction.

(3)

The European Central Bank (ECB) decided a year ago to hold this week's monetary-policy meeting in Barcelona, but the timing turned out to be perfect. Spain is in the crosshairs of the markets, not least because of budgetary overruns by regional governments such as Catalonia's. And the contrasting economic fortunes of beaten-up Spain, where the jobless rate has reached 24%, and resilient Germany, where it is below 6%, exemplify the difficulty of finding the right monetary policy in a currency union of 17 members.

(4)

It was many hundreds of years before any further significant observations were made about the phenomenon of static electricity. Then it was discovered that many other materials besides amber could be charged by rubbing, which produced friction. A more important discovery was that there were two kinds of electrical charges.

These two kinds of charges were called positive and negative. A positive charge was indicated by a plus sign (+) and a negative charge by a minus sign (−). These symbols are still in universal use today. It was also discovered that like charges—two positive charges or two negative charges—repelled each other, whereas unlike charges—a positive and a negative charge—attracted each other.

(5)

Between 1993 and 1997 the researchers questioned 20, 000 healthy British men and women about their lifestyles. They also tested every participant's blood to measure vitamin C intake, an indicator of how much fruit and vegetables people ate. Then they assigned the participant—aged 45 – 79—a score of between 0 and 4, giving one point for each of the healthy behaviors. After allowing for age and other factors that could affect the likelihood of dying, the researchers determined people with a score of 0 were four times as likely to have died, particularly from cardiovascular disease. The researchers, who tracked deaths among the participants until 2006, also said a person with a health score of 0 had the same risk of dying as someone with a health score of 4 who was 14 years older. The lifestyle change with the biggest benefit was giving up smoking, which led to an 80 percent improvement in health, the study found. This was followed by eating fruits and vegetables. Moderate drinking and keeping active brought the same benefits, Key-Tee Khaw and colleagues at the University of Cambridge and Medical Research Council said. "Armed with this information, public-health officials should now be in a better position to encourage behavior changes likely to improve the health of middle-aged and older people," the researchers wrote.

三、拓展阅读和翻译练习

将以下两篇英文科技语篇翻译成对等的中文科技语篇。

篇章1

An integrated circuit, commonly referred to as an IC, is a microscopic array of electronic circuits and components that has been diffused or implanted onto the surface of a single crystal, or chip, of semiconducting material such as silicon. It is called an integrated circuit because the components, circuits, and base material are all made together, or integrated, out of a single piece of silicon, as opposed to a discrete circuit in which the components are made separately from different materials and assembled later. ICs range in complexity from simple logic modules and amplifiers to complete microcomputers containing millions of elements.

The impact of integrated circuits on our lives has been enormous. ICs have become the principal components of almost all electronic devices. These miniature circuits have demonstrated low cost, high reliability, low power requirements, and high processing speed compared to the vacuum tubes and transistors which proceeded them. Integrated circuit microcomputers are now used as controllers in equipment such as machine tools, vehicle operating systems, and other applications where hydraulic, pneumatic, or mechanical controls were previously used. Because IC microcomputers are smaller and more versatile than previous control mechanisms, they allow the equipment to respond to a wider range of input and produce a wider range of output. They can also be reprogrammed without having to redesign the control circuitry. Integrated circuit microcomputers are so inexpensive they are even found in children's electronic toys.

The first integrated circuits were created in the late 1950s in response to a demand from the military for miniaturized electronics to be used in missile control systems. At the time, transistors and printed circuit boards were the state-of-the-heart electronic technology. Although transistors made many new electronic applications possible, engineers were still unable to make a small enough package for the large number of components and circuits required in complex devices like sophisticated control systems and handheld programmable calculators. Several companies were in competition to produce a breakthrough in miniaturized electronics, and their development efforts were so close that there is some question as to which company actually produce the first IC. In fact, when the integrated circuit was finally patented in 1959, the patent was awarded jointly to two individuals working separately at two different companies.

After the invention of the IC in 1959, the number of components and circuits that could be incorporated into a single chip double every year for several years. The first integrated circuits contained only up to a dozen components. The process that produce these early ICs was known as small scale integration, or SSL. By the mid-1960s, medium scale integration, MSL, produced ICs with hundreds of components. This was followed by large scale integration techniques, or LSI, which produced ICs with thousands of components and made the first microcomputers possible.

The first microcomputer chip, often called a microprocessor, was developed by Intel Corporation in 1969. It went into commercial production in 1971 as the Intel 4004. Intel introduced their 8088 chip in 1979, followed by the Intel 80286, 80386 and 80486. In the late 1980s and early 1990s, the designations 286. 386 and 486were well known to computer users as reflecting increasing levels of computing power and speed. Intel's Pentium chip is the latest in this series and reflects an even higher level.

Source: https://zhidao.baidu.com/question/24512963.html

篇章 2

Genetically modified foods—Feed the World?

If you want to spark a heated debate at a dinner party, bring up the topic of genetically modified foods. For many people, the concept of genetically altered, high-tech crop production raises all kinds of environmental, health, safety and ethical questions. Particularly in countries with long agrarian traditions—and vocal green lobbies—the idea seems against nature.

In fact, genetically modified foods are already very much a part of our lives. A third of the corn and more than half the soybeans and cotton grown in the US last year were the product of biotechnology, according to the Department of Agriculture. More than 65 million acres of genetically modified crops will be planted in the US this year. The genetic is out of the bottle.

Yet there are clearly some very real issues that need to be resolved. Like any new product entering the food chain, genetically modified foods must be subjected to rigorous testing. In wealthy countries, the debate about biotech is tempered by the fact that we have a rich array of foods to choose from—and a supply that far exceeds our needs. In developing countries desperate to feed fast-growing and underfed populations; the issue is simpler and much more urgent: Do the benefits of biotech outweigh the risks?

The statistics on population growth and hunger are disturbing. Last year the world's population reached 6 billion. And by 2050, the UN estimates, it will probably near 9 billion. Almost all that growth will occur in developing countries. At the same time, the world's available cultivable land per person is declining. Arable land has declined steadily since 1960 and will decrease by half over the next 50 years, according to the International Service for the Acquisition of Agri-Biotech Applications (ISAAA).

The UN estimates that nearly 800 million people around the world are undernourished. The effects are devastating. About 400 million women of childbearing age are iron deficient, which means their babies are exposed to various birth defects. As many as 100 million children suffer from vitamin A deficiency, a leading cause of blindness. Tens of millions of people suffer from other major ailments and nutritional deficiencies caused by lack of food.

How can biotech help? Biotechnologists have developed genetically modified rice that is fortified with beta-carotene—which the body converts into vitamin A—and additional iron, and they are working on other kinds of nutritionally improved crops. Biotech can also improve farming productivity in places where food shortages are caused by crop damage attribution to pests, drought, poor soil and crop viruses, bacteria or fungi.

Damage caused by pests is incredible. The European corn borer, for example, destroys 40 million tons of the world's corn crop annually, about 7% of the total. Incorporating pest-resistant genes into seeds can help restore the balance. In trials of pest-resistant cotton in Africa, yields have increased significantly. So far, fears that genetically modified, pest-resistant crops might kill good insects as well as bad appear unfounded.

Viruses often cause massive failure in staple crops in developing countries. Two years age, Africa lost more than half its cassava crop—a key source of calories—to the mosaic virus. Genetically modified, virus-resistant crops can reduce that damage, as can drought-tolerant seeds in regions where water shortages limit the amount of land under cultivation. Biotech can also help solve the problem of soil that contains excess aluminum, which can damage roots and cause many staple-crop failures. A gene that helps neutralize aluminum toxicity in rice has been identified.

Many scientists believe biotech could raise overall crop productivity in developing countries as much as 25% and help prevent the loss of those crops after they are harvested.

Yet for all that promise, biotech is far from being the whole answer. In developing countries, lost crops are only one cause of hunger. Poverty plays the largest role. Today more than 1 billion people around the globe live on less than $ 1 a day. Making genetically modified crops available will not reduce hunger if farmers cannot afford to grow them or if the local population cannot afford to buy the food those farmers produce.

Nor can biotech overcome the challenge of distributing food in developing countries. Taken as a whole, the world produces enough food to feed everyone—but much of it is simply in the wrong place. Especially in countries with undeveloped transport infrastructures, geography restricts food availability as dramatically as genetics promises to improve it.

Biotech has its own "distribution" problems. Private-sector biotech companies in the rich countries carry out much of the leading-edge research on genetically modified crops. Their products are often too costly for poor farmers in the developing world, and many of those products won't even reach the regions where they are most needed. Biotech firms have a strong financial incentive to target rich markets first in order to help them rapidly recoup the high costs of product development. But some of these companies are responding to needs of poor countries. A London-based company, for example, has announced that it will share with developing countries technology needed to produce vitamin-enriched "golden rice".

More and more biotech research is being carried out in developing countries. But to increase the impact of genetic research on the food production of those countries, there is a need for better collaboration between government agencies—both local and in developed countries—and private biotech firms. The ISAAA, for example, is successfully partnering with the US Agency for International Development, local researches and private biotech companies to find and deliver biotech solutions for farmers in developing countries.

Will "Frankenfoods" feed the world? Biotech is not a panacea, but it does promise to transform agriculture in many developing countries. If that promise is not fulfilled, the real losers will be their people, who could suffer for years to come.

Source: https: //www. docin. com/p-2061639369. html

第8章　不同类型英文科技语篇翻译实例

科技英语是英语的科技语体，可以泛指一切论及科学和技术的书面语和口语，包括科技著述、论文、报告、各类科技情报、科技实用手段的结构描述和操作描述，有关科技问题的会谈、会议，以及有关科技的影片、录像等有声资料的解说词，等等。下文将讨论常见的不同类型英文科技语篇的翻译。

第1节　科技新闻的翻译

英语科技新闻指"对最近发生的科技及其有关事实所进行的科学性、知识性报道。"其语言是科技语言与新闻语言的结合，但又不是简单的相加，而是在二者的基础上形成了自己的特点。总之，科技新闻是新闻与科技的结合，既具有新闻报道的共性，也有科技文体的特性。科技新闻报道也同样由标题（headline）、导语（lead）、正文（body）三个部分组成。导语一般都在正文的第一段，其目的是将人物、事件、时间、场所、原因、方法、结果均交待清楚，其要点和结论也通常在第一段概括提出，而详情则在导语后面的段落进一步说明。由于英语的新闻标题有其独特的语法体系，是英语科技新闻阅读理解及翻译的关键，以下将讨论新闻标题的特点及翻译。

1.1　科技新闻标题的句法特点及翻译

标题的功能是引起读者的阅读兴趣以及告知新闻的主要内容。英语新闻的标题具有以下特点：

（1）常省略不必要的虚词与系动词。

英语新闻标题中，冠词、连词、介词、系动词、助动词等常予以省略。翻译时则应将所省略成分补全，再斟酌翻译。例如：

River Still Polluted After Clean-up Efforts

分析：

在这个标题中，系动词 is 被省略。

译文：

河流虽经清理，依旧污染如故

No Survivors in Gulf Air Crash

分析：

在这个标题中省略了 there be 结构及定冠词 the，完整的标题应是（There are）No

Survivors in (the) Gulf Air Crash

译文：

海湾空难中无人幸存

Rail Chaos Getting Worse

分析：

在这个标题中省略了定冠词 the 及系动词 is，完整的标题应该是 (The) Rail Chaos (Is) Getting Worse

译文：

铁路系统的混乱局面愈演愈烈

（2）大量使用现在时表示过去发生的事情，不定式表示将要发生的事情。

标题中动词时态用法大大简化，为了使读者感到读到的是“新闻”而不是“旧闻”，几乎都用一般现在时表述新闻，这是标题的另一个重要特点。在翻译的时候，应该根据正文所描述的实际情况予以翻译。例如：

EU **Plans** to Boost France's Recovery

分析：

Plans 其实是已经发生的事情，因此，译文中应该有所体现。

译文：

欧盟已计划促进法国的经济恢复

Swedish Oil Deliveries **Halt** as Strike Spreads

分析：

halt 在此则指即将发生的事情。

译文：

由于罢工的蔓延，瑞典的原油交付将停止

（3）大量使用缩略词以及小词。

为节省篇幅，标题中普遍使用缩略词，例如，常用机构的首字母缩略词简称，而不是全名及小词（midget words）。这就要求译者必须掌握一定数量报刊中出现频率较高的政治、军事、经济、文化教育等重要机构的简称。一些拼写较长的词常用其截短的形式代替，例如：earthquake，refrigerator 分别常用 quake 和 fridge 来代替，单音节词常代替同义的多音节词，例如：aid 常代替 help，assist，而 back 常代替 support。例如：

OPEC to Raise Production

分析：

(OPEC = Organization of the Petroleum Exporting Countries 石油输出国组织)

译文：

石油输出国组织将增大产量

The Great Superpower Spy War: KGB vs CIA

分析：

KGB = Komitet Gosudarstvennoi Bezopastnosti (克格勃，苏联国家安全委员会)

vs = versus

CIA = Central Intelligence Agency

译文：

超级大国的间谍战：克格勃对中情局

No Hope for 118 Crew of Russian Sub

分析：

sub = submarine

译文：

俄罗斯潜艇118名船员生存希望渺茫

Put the Sci Back in Sci-fi

分析：

sci = science

fi = fiction

译文：

让科学重返科幻小说

Man **Claims** Ghost Sighting

分析：

Claim = declare

译文：

男子声称看见鬼

Peace **Drive** Succeeds

分析：

Drive = effort/campaign

译文：

和平努力取得成功

1.2 科技新闻报道的写作风格与翻译

科技新闻报道具有五大特点，即准确、新颖、通俗、精练、生动。准确（accuracy），是因为新闻的生命在于真实，科技新闻尤其如此；新颖（originality），是因为科技新闻要突出反映新成果、新技术，所以应多在“新”字上着眼；通俗（popularity），是因为其目标读者是普通读者，而非专业人士，因此，报道必须在准确的前提下做到通俗易懂，所以记者会尽量避免过于专业化的词语，如非用不可时，也会用准确通俗简短的语言加以解释；精练（succinctness），是因为科技报道不是科研报告或者学术讲座，无须反映研究过程或者细节，只要酌情报道最重要、最精华的事实，即科技成果及其成功的关键与意义；生动（vividness）是指科技新闻应该是有趣的、吸引人的。鉴于以上这些特点，科技新闻在写作风格上形成了以下的特点：

（1）为求通俗，慎用术语。

科技新闻报道的对象是科学技术及相关事实，但新闻的受众却是一般大众，因此，必须考虑受众的接受能力，以通俗易懂的方式来传达科技信息。在英语科技新闻中，记者通常会用普通词汇代替专业术语词汇，以达到通俗易懂的目的。那么在翻译时，译者也应准确理解某个通俗词语的确切意思，切勿望文生义、字对字硬译，同时要传达出通俗易懂这一特点来。例如：

The **white blood cells** of the immune system are always bumping into cancer cells. They should attack cancer cells as foreign bodies and destroy them.

分析：

在这个例句中，white blood cells 指"白血球、白细胞"，等同于医学专业术语 leukocyte, 切勿字对字翻译成"白的血细胞"。

译文：

免疫系统的白血球总会撞上癌细胞。它们应该会将癌细胞当作异物而予以摧毁。

One of Hwang's students said she and another student had donated **eggs** for use in the lab's cloning experiments.

分析：

在这个例句中，eggs 等于 egg cell 意思是"卵子、卵细胞"，其生物学及医学的专业术语则是 ovum, 因此，译者在翻译时应该根据科技新闻写作力求通俗这个特点，正确理解 egg 一词的意思，并恰当地转换。

译文：

黄的一个学生说，她和另一个学生已经捐献了一些卵子用于该实验室的克隆实验。

（2）为求生动，活用数据。

科学技术讲究严谨、精确、量化，多数科技成果都是对实验数据进行分析后得到的，所以，报道科技事件和动态大多会涉及数字信息。但数字是抽象的，往往要在特定语境中才能显示出意义。如果只是简单罗列相关数字，而不能揭示其后面的意义，受众看了会一头雾水。因此，记者在写科技新闻时常会对数字进行灵活处理，以便让枯燥抽象的数字生动起来，成为被受众接受的有意义的信息。译者在翻译这些数字时要准确地以中文相应的方式翻译出来。例如：

Nie Haisheng and Fei Junlong circled Earth for five days aboard the Shenzhou6 capsule, travelling 2 million miles in 115 hours, 32 minutes. China's first manned mission was in 2003, when astronaut Yang Liwei orbited 211/2hours.

分析：

这是关于中国第二次发射载人飞船神舟 6 号的新闻报道。"聂海胜与费俊龙乘坐神舟 6 号飞船绕地球飞行了 115 小时又 32 分钟"，为了让受众对这个数字以及其背后的意义有深刻的感受，新闻后面对比了第一次载人飞船在太空飞行了 21.5 个小时，通过在太空停留时间的前后对比，使受众对中国航天事业的发展有了更深刻的认识。

译文：

聂海胜与费俊龙乘坐神舟 6 号飞船绕地球飞行了五天，在 115 小时又 32 分钟内飞行了2百万英里的距离。中国第一次载人航天飞行任务是在2003年进行的，当时，航天员杨利伟绕轨道飞行了21.5小时。

科技新闻中的数字比较抽象，有时为了让受众留下深刻印象，记者会通过换算，以比例或者倍数的形式对数字加以诠释，从而将事情讲得更清楚透彻。这时，译者要注意中英文倍数表达的差异，准确翻译倍数及百分比。例如：

Sonnet Technologies released a line of high-capacity 3.2-volt batteries for all iPods...will play up to 78 percent longer than the original iPod batteries.

分析：

在该例子中，新电池的容量为3.2伏，78%是指电池净增的播放时间，可以译为“比旧电池播放的时间延长了78%。”

译文：

索内科技公司为所有苹果公司的音乐播放器推出一系列3.2伏的高容量电池……这将使播放时间比原来的电池延长了78%。

He said Chinese space officials want to study the possibility of making rockets with the capacity of carrying spacecraft weighing 27.5 tons—three times the capacity of their existing rockets.

分析：

在这一个例子中，记者将27.5吨换算成一个倍数，即，是现有火箭承载能力的3倍，以对这个数字进行诠释，从而使读者更好地感受到中国航天科技的巨大进步。

译文：

他说，中国航天部的官员希望研究制造出承载能力为27.5吨的火箭的可能性，这个承载能力是现有火箭的3倍。

（3）广泛使用缩略语。

科技新闻中涉及的概念或者术语会在报道中反复出现，为了节约空间，同时使读者便于阅读，广泛使用缩略语。例如，在《纽约时报》一篇题为“Screening for Abnormal Embryos Offers Couples Hope after Heartbreak”中反复使用了P. G. D. 这个缩略语，在第一次使用时作者对其进行了解释：“Preimplantation genetic diagnosis, referred to as P. G. D. ”（胚胎植入前遗传学诊断）。译者在翻译过程中需要熟悉这些缩略语表达的意义，根据情况酌情翻译。

一般新闻中使用的缩略语可直接使用单词首写字母缩略形式，不需另外进行解释说明，而且读者接受度很高的缩略语可以采用零翻译方式处理。例如：

That's extremely pricey for a laptop, and it doesn't even include the $350 external module you'll need to play or record DVD's and CD's. Then again, the Flybook's makers have adopted an iPod dish marketing strategy, hoping to sell the thing as jewelry.

分析：

在此例中，DVD的全称是Digital Videodisc（数字视盘），CD的全称是Compact Disc（光盘），二者都是受众接受度很高的缩略语。其意义流传甚广，因此，已经无须解释。在翻译时这两个缩略语可以采用零翻译方式处理，即在译文中保持原形。

译文：

这对笔记本电脑而言价格过高，而且它甚至不包括价格350美元、用来播放或者录制DVD或者CD的外置配件。再者，飞书阅读器的制造者们已采用了iPod的市场销售策略，希望能借此将产品卖出好价钱。

但如果新闻中出现的是新生词汇，或者是不为大众所熟悉的概念或者术语，一般记者会在文中予以解释，同样，译者也需要翻译出这个缩略语的意义来。例如：

The aptly-named FASTER study (for first-and second-trimester evaluation of risks) published in the "New England Journal of Medicine"...

分析：

在本例中记者采用括号里写出全称的方式对FASTER这个缩略语予以说明，既为读者扫清了阅读理解的障碍，也便于在下文中继续使用该缩略语。作为译者，也应该考虑到译文读者对此缩略语的接受程度，将此解释翻译出来，即，"妊娠初期和中期的风险评估"。这样，当该缩略语在新闻中再次出现时，则可以采用零翻译方式处理。

译文：

恰如其名的FASTER快捷研究，即对妊娠初期和中期的风险评估的研究，已在《新英格兰医学期刊》公开发表……

(FASTER既是一个首字母拼音词，也是记者用到的双关语，取其"更快的"意义与前置修饰语aptly-named搭配使用，因此，对此缩略语采用了综合处理的方法，即零翻译+意译+注释)。

(4) 为求客观，大量使用第三人称主语及被动语态。

科技新闻侧重叙事推理，强调客观准确，因此，常常以第三人称的叙述角度报道新闻，并常使用被动语态。这一特点与一般科技文体的特点是相符的，在翻译时译者要格外注意语态的转换。例如：

Alcohol dependence, also known as alcoholism, is defined as impaired control over drinking that disrupt work, school or home life.

分析：

在本例中，运用被动语态及含有被动意义的后置定语以精确客观地界定所报道的事实。由于中文更习惯使用主动语态，译者在翻译时应该注意恰当转换语态。

译文：

酒精依赖，又称酗酒，其定义是对饮酒的控制力受损，以至于破坏（正常的）工作、学习及家庭生活。

(5) 为求文体活泼，使用比喻手法。

虽然科技新闻通常的基调是比较严肃的，但为了增添活泼感，以吸引读者的眼球，也常采用比喻的手法以使报道生动形象。这点在一般科技文体中比较少见，译者在翻译时应正确理解报道中某些词语的比喻用义，再忠实地翻译出来，使译文读者也能获得与原文读者同样或者类似的感受。例如：

An unexpected benefit of the effluent treatment plant is that it has acted as a "**policeman**" to the metallurgical operations.

It'll be useless to select the proper mechanicals if the drive system turns out to be a "**bottleneck**".

分析：

以上两个例子分别用到了明喻（例1）与暗喻（例2）的修辞手法，使原本枯燥的信息跃然纸上。在例1中，policeman(警察）一词比喻monitor(监测器)，生动形象而且易懂；例2中，bottleneck相当于汉语中的“瓶颈”，其比喻意义为“inoperative”，指“无效的、不起作用的”，从而突出了驱动系统的关键作用。

译文：

污水处理厂的一个意外的好处是，它可以充当冶金作业的监测器。

如果驱动系统最终起不了作用，就是选择了适当的机械设备也没有用。

1.3 翻译实例

英语科技新闻有机地融合了新闻语言及科技语言的特点，表现出独特的写作风格。译者在翻译的时候，务必要正确理解原文，灵活采用适当技巧如实传达原文的信息，以期达到与其同等的表达效果。例如：

Freshwater Turtle Endangered

More than 1/3 of the 280 known species of freshwater turtles currently face extinction, according to a new analysis by Conservation International (CI), BBC reported. Capture of turtles for their purported medicinal properties and to supply a lucrative pet trade are the key drivers behind the fall in numbers. Habitat loss as a result of river-damming for hydro-electricity is another major concern. The species in deepest trouble is the red river giant softshell turtle, with only 4 individuals remaining alive in the world. The red-crowned river turtle and the Myanmar river turtle are also in great danger.

分析：

在标题Freshwater Turtle Endangered中，省略了系动词，新闻开门见山引出报道的对象，即淡水龟，然后才介绍新闻来源。根据中文的表述习惯，应该颠倒语序，先交代出处，再陈述事实。此处还要注意两个机构的译名，一个采用了全称，一个是缩略语，即Conservation International和BBC。之后分析了淡水龟濒临灭绝的两个最主要的原因，以及最受影响的几个品种。除第一句颠倒语序以外，其余部分基本上可以按照原文语序顺译出来。

译文：

淡水龟濒临灭绝

据英国广播公司报道，保护国际基金会（CI)的一项最新研究表明，现存已知的280种淡水乌龟中有超过1/3品种正濒临灭绝。淡水龟大量减少的主要原因是人们认为其含有药用价值，因而大肆捕捉，或将其用于利润丰厚的宠物交易。此外，因水电工程而修建水坝导致淡水龟的栖地流失，也是人们关注的另一大原因。目前濒危程度最高的品种是斑鳖，全球仅剩4只存活。红冠棱背龟和缅甸棱背龟也同样面临巨大的危险。

Underground Test in U. S.

NEW YORK, Thurs., APP——A nuclear device as powerful as 90,000 tons of TNT was detonated a mile beneath the Colorado Rockies today in an experiment aimed at loosening vast stores of natural gas.

The explosion of the device's three 30-kiloton atomic mechanism went smoothly.

It was the largest nuclear blast so far in the Atomic Energy Commission project to stimulate massive stores of natural gas locked in rock formations.

The AEC said only background levels of radiation normally found in the environment was recorded at a monitoring station at the blast site.

The blast shock the ground at an observation area 11 miles from the site and street lights 90 miles away in Montrose, Colorado.

Residents of Rifle, about 30 miles from the site of the detonation, said they felt nothing.

The AEC had predicted that the buildings in the town would shake.

The National Earthquake Centre 180 miles away at Boulder recorded the blast 40 seconds after it was set off.

Center spokesman Jim Lander said seismographs recorded the jolt at 5.3 on the Richter Scale, which has no top limit.

Mr. Lander said the tremor would be classed as a "locally severe earthquake which would cause damage."

译文：

美国的一次地下核试验

[美联社纽约星期四电] 今天，在科罗拉多落基山区地下一英里处，爆炸了一个能量为九万吨的三硝基甲苯炸药的核装置。这是一次以释放储量丰富的天然气为目的的实验。

该装置三个三万吨级原子结构的爆炸进行顺利。

到目前为止，这是原子能委员会开发蕴藏在岩层中大量天然气工程的一次最大的核爆炸。

原子能委员会说，爆炸现场监测站记录到的只是当地环境所常有的基本辐射强度。

这次爆炸震动了离现场11英里处观察区的地面，和90英里外科罗拉多州蒙特罗斯城的街灯。

距爆炸现场大约30英里的莱福镇居民说，他们毫无感觉。

原子能委员会曾预计该镇的建筑物会发生震动。

180英里外的布尔德国家地震中心四十秒后记录到这次爆炸。

该中心发言人金·兰德说，地震仪记录的震动强度为里氏5.3级，这种震动没有强度上限。

兰德先生说，可把该震动列为"有破坏性的局部严重地震"的类别。

第2节 科普语篇的翻译

2.1 科普语篇的文体特点

科普文体是科技文体的一部分，方梦之在《科技翻译：科学与艺术同存》一文中对科技文体就文体的正式程度（formality）做了三种划分：科学论文（scientific paper）、科普文章（popular science article）、技术文本（technical prose document）。科学论文是专家写给专家看的（expert-to-expert），正式程度最高；科普文章是内行写给外行看的（scientist journalist-to-lay-person writing），科普作者要把科学道理说清楚，要极尽其运用文学修辞手段之能事。相对而言，科普文章的正式程度较低。而技术文本，包罗是很广的，它既可是专家写给专家看的（expert- to-expert），也可是专家写给外行看的（expert-to-layperson），像项目申请书、可行性报告、产品说明书、维修手册、技术协议书等均是技术文本。显而易见，“科普文章”远比“科学论文”和“技术文本”离普通读者更近。

科普文体是科技文体的一个支派，是文学和科学相结合的写作体裁。因为科普文章的目的是普及科学技术知识，所以除科学性和文学性外，一般还具有通俗性和趣味性的特点。例如：

It (wood frog) spends its winters interned in subzero sleep, its tissues steel-rigid, and revives in the spring raring to go. It's the Rip Van Winkle of the animal world.

分析：

作者这里用修辞手法把林蛙比作了“瑞普·凡·温克尔”，温克尔是美国作家华盛顿·欧文小说中的主人公，因酒后昏然入睡，一觉睡了20年，醒来时自己已经成了老人，世界也面目全非了。这是非常形象生动的描写，所以，宣传科学并不是枯燥的概念和干巴巴的学术名词的简单堆砌。有时，生动形象的文学语言更有助于科学的传播。因为考虑到译文读者未必知道欧文这个故事中的人物，为了传达出典故的内涵，而增译“睡神”一词。这样，即便读者没有看过欧文的这个小说，也能领悟到这个比喻的意义。

译文：

在冬天，它（林蛙）的体温降到零度以下，处于休眠状态，器官组织没有任何活动，而到了春天，它就会复活。它是动物界的睡神“瑞普·凡·温克尔”。

由于简明英语运动（Plain English Campaign）在很多国家蓬勃发展……它反对使用别人不懂的行话、官样文章和其他含糊不清的语言，因此，自然、简洁、朴实的文风越来越受到人们的喜爱。甚至可以说，科普文体和普通英语文体并无多大差异，差别只在于科普文章说的是科学的内容，科学技术词汇更多一些而已。例如：

Scientists used to believe the virus jumped to humans after incubating in pigs, which can be infected by both bird and human viruses. But the London researchers have shown a few small changes in the shape of a surface protein were all it took to enable the bird version of Spanish flu to bind onto human cells. (*Discover May* 2004: 8)

译文：

过去，科学家认为这种病毒是先潜伏在猪的体内，然后才进入人体的，禽类和人类的病毒都能使猪受到感染。但是，英国的研究人员表示，他们发现了病毒表面蛋白质形状上的几处细微变化，正是这些变化使西班牙流感的禽类病毒得以附着在人的细胞上。

2.2 科普语篇的翻译

公众距离最近、接触最多的是科普文章，从报章小品到科普杂志，从日常生活的科学技术到对国外科技新动态的介绍均属科普范畴。科普文章的翻译应该是纯科技文本翻译方法与文学翻译方法的结合，既要体现科技内容的科学严谨，又要使人喜闻乐见，这样才能真正达到普及科学知识的作用。因为科普文章的广大读者不可能具备方方面面的科技知识，只有通俗易懂，晓畅明白，同时又能忠实传达原文艺术形象的译文才会受到广大读者的喜爱，也才能最大限度地传播科学技术。所以，科普翻译的原则应该是准确达意，通俗易懂，兼具文采。如果译文抹杀了原文的艺术性，只有干巴巴的术语和数据，科学普及将势必成为一句空话。但应该注意的是，在科普翻译中，译文的艺术性应服从于译文的正确性，即不能为了美学的目的而有损于原文的科学性。例如：

Because one of the abiding aims of civilization is to make life safer for people, Austad says, the trend toward a longer life span will continue, and the luxury of long life, afforded by a civilized lifestyle, will eventually become encrypted in our DNA. Nurture becomes nature; culture dictates biological destiny.

分析：

原文用词准确、严谨，长短句交错使用，各司其职，很好地表达了作者的说理意图。译文通过精心的措辞，充分运用汉语灵活多样的表达手段，也基本达到了准确达意、通俗易懂的目的。由于科普作品的目的是普及科学知识，传播科技方法，多数作者都愿意用丰富多彩的语言（包括各种修辞方法）来传播信息和表达思想。读者的美学要求决定了科普作品中文学色彩的必然性。扎实的外语和中文功底是翻译的基础，而一定的科学常识和文学修养则是一个合格的科普译者的必备条件。只有这样，才能保证译文的正确性和可读性。

译文：

奥斯塔德说，因为文明的一个永恒目的是使人们生活得更安全，所以人类长寿的趋势将会持续。文明的生活方式帮助我们实现了长寿的奢望，人类长寿的理想最终可以通过控制人类的 DNA 排列而得到充分实现。后天培养成为自然的过程，而文化熏陶将主宰生物的命运。

又例如：

Michael Faraday

During the first half of the 19th century scientists began to acquire a better understanding of electricity. We owe much to people like Volta, Ohm and Ampere whose names are associated with the various units of electrical measurement, and to Michael Faraday whose later discoveries laid the foundation for our modern electrical industry.

Faraday was one of the world's greatest experimenters and he set himself the task of finding out more about light, heat, electricity and magnetism. He had the idea that if electricity flowing through a wire it could create a magnetic field around it (a fact discovered by Ampere) then the field around a magnetic might be used to produce electricity. His experiments with two magnets, an iron core and a coil of wire proved his idea to be correct.

From these early experiments in electro-magnetism came eventually the development of the electric generator which produces electricity for our homes and factories. They also resulted in the invention of the electric motor to drive machinery and provide power for electric trams and trains. The electric telegraph, the telephone and many electrical facilities owe their development to the original work of Michael Faraday.

分析：

介绍科学家的文章和书籍是科普读物的一个重要组成部分。这篇短文就是对著名英国物理学家法拉第的简略介绍，主要提到他对电力工业和电器发明所做出的巨大贡献。文中虽提到他曾用“两块磁体、一个铁心和一组线圈”进行“试验”，但并未阐述发电的原理和他是如何作试验的，因为对科学家的介绍虽涉及科学内容，但如人物传记一样，着重在介绍人物，特别是他在科学发展方面做出的功绩。如过多涉及科学的专门知识不仅使普通读者难以读懂，也会喧宾夺主，反将科学家本人置于次要地位。故文章作者只用最基本的科技词汇，如 light, heat, electricity, magnetism, magnetic field, iron core, coil of wire, generator, motor 等。这些都在一般读者的知识范围之内，不会带来理解上的困难。此外文章的风格比较正式，选用的词汇短语也比较正规，如 associated with, laid the foundation for, resulted in, owe…to 等。译文也应用相同的较正式的汉语词汇、短语来翻译这些基本科技词汇和正规文体的表达方式，如“光、热、电、磁力、磁场、铁心、线圈、发电机、电动机”，以及“联系在一起”“奠定基础”“导致”“归功于”等，较准确地体现了原文的风格。

译文：

米迦勒·法拉第

19 世纪前半期，科学家们开始对电有了更深入的了解。这都是伏特、欧姆、安培这些人的功劳，他们的名字已经和电的各种测量单位联系在一起。也应归功于米迦勒·法拉第，他的发现奠定了现代电力工业的基础。

法拉第是世界上最伟大的实验物理学家之一。他的目标是进一步弄清楚光、热、电和磁力的特性。他认为既然电在通过电线时会在电线周围形成磁场（安培发现了这一点），那么，磁体周围的磁场就一定可以用来产生电。他用两块磁体、一个铁心和一组线圈进行试验，结果证明他的想法是正确的。

后来为家庭和工厂生产电力的发电机最终就由这些早期的电磁实验发展而来。这些实验也导致了能开动机器，和为电车与电气火车提供动力的电动机的发明。电报、电话以及其他许多电器设备的发明也都应该归功于米迦勒·法拉第的创造性劳动。

第3节 科幻小说的翻译

科幻小说从文体上讲，属于文学类别，其主要任务是供读者消遣和娱乐，激发他们对科学的兴趣。科幻小说最大的特点是文学性与科学性并存，既具有一般小说的共同性，又具有内容上的科学性，甚至有的科幻小说还有十分突出的前瞻性。但由于科幻小说属于通俗小说的范畴，针对的读者对象是一般大众，其阅读目的是娱乐和消遣，他们不希望做出太多的阅读努力。因此，科幻小说的翻译标准必然与一般严肃文学的翻译标准也有所区别，这就要求译文首先要通俗易懂。

科幻小说也会涉及科学技术问题，优秀的科幻小说往往是在现实科学的基础上，或描述未来的科学，或再现过去的科学发现。这些科学的表述，必须在译文中得到准确的反映。在论述科普作品的翻译标准时，把准确性放在第一位，文学性放在第二位。但是，科幻小说作为一种文学的样式，翻译标准就必须把文学性放在第一位，科学性退而居次。根据以上论述，可以把科幻小说的翻译标准定为：第一突出文学性；第二是科学性；第三是通俗性。

作为小说，作品中必然会有文学性较强的部分，比如人物刻画、环境的描写、情节的建构等。因此，在翻译文学性较强的段落时，译者要用对等的文字表现其文学性；又由于是科幻类别的小说，其内容必然涉及科学技术，因此，在翻译科学性较强的段落时，译者应该要注意表达的科学性，即强调相关内容的准确性。考虑到读者娱乐消遣的目的，在遣词造句时，译者要注意通俗易懂，也就是对于术语尽量用通俗的词语来表达，不要专业化过强；在做文学描述的时候，也同样用现代读者普遍接受的语言来描述，不要过于文雅，以免曲高而和者寡。例如：

"Clearly," the Time Traveler proceeded, "any real body must have extension in FOUR directions: it must have Length, Breadth, Thickness, and—Duration. But through a natural infirmity of the flesh, which I will explain to you in a moment, we incline to overlook this fact. There are really four dimensions, three which we call the three planes of Space, and a fourth, Time. There is, however, a tendency to draw an unreal distinction between the former three dimensions and the latter, because it happens that our consciousness moves intermittently in one direction along the latter from the beginning to the end of our lives."...

"Now, it is very remarkable that this is so extensively overlooked," continued the Time Traveler, with a slight accession of cheerfulness. "Really this is what is meant by the Fourth Dimension, though some people who talk about the Fourth Dimension do not know they mean it. It is only another way of looking at Time. THERE IS NO DIFFERENCE BETWEEN TIME AND ANY OF THE THREE DIMENSIONS OF SPACE EXCEPT THAT OUR CONSCIOUSNESS MOVES ALONG IT. But some foolish people have got hold of the wrong side of that idea. You have all heard what they have to say about this Fourth Dimension?"

分析：

上述段落引自英国作家赫伯特·乔治·威尔斯创作的中篇小说《时间机器》，该作讲述时间旅行者发明了一种机器，能够在时间纬度上任意驰骋于过去和未来。当他乘着机器来到公元802701年时，展现在他面前的是一幅奇异恐怖的景象。人类分化为两个种族：爱洛伊人和莫洛克人。前一种人长得精致美丽，但失去了劳动能力。后一种人则面目狰狞，终日劳动，过惯了地下潮湿阴暗的生活。他们养肥了爱洛伊人，到了晚上便四处捕食他们。

上述段落是作品开头部分，时间旅行者在向一些人介绍他的发明——时间机器的构思。他在尽可能用简单明了的话语使一般人能明白其发明中蕴含的科学原理，其中并没有特别艰深的科学道理，又因旅行者是在口头解释，翻译时还要注意一些普通词汇用作术语的情况，例如，“dimension”“plane”，以及文中出现的一些替代词语诸如，“the former”“the latter”，代词“it”的转换，必要时应该将其还原成其所指代的事物，以使意义明晰。最后，还要注意语言的通俗性。

译文：

“显然”，时间旅行者继续说道，“任何一个实在的物体都必须向四个方向伸展：它必须有长度、宽度、高度和时间持续度。但由于人类肉体凡胎天生的缺陷，我们往往忽视这个事实，这点我待会儿再解释。实际上有四个维度，其中三个维度是我们称作空间的三个平面，第四维就是时间。然而，人们现在总喜欢在前三者和后者之间划上一条实际并不存在的区分线，因为正巧我们的意识从生命的开始到结束正是沿着时间的同一方向断断续续朝前运动的。”……

“是啊，许多人都忽视了这一点，真是不可思议。”时间旅行者继续说道，他的兴致更浓了。“实际上，这就是第四维的内涵，虽然有些人谈论第四维时并不知道他们指的就是这个意思。这其实只是看待时间的另一种方式。时间和空间三维中的任何一维之间并没有什么不同，区别只是我们的意识是沿着时间向前运动的，可有些笨蛋把这个观点的意思搞颠倒了。你们听过他们有关第四维的高见吗？”

I grieved to think how brief the dream of the human intellect had been. It had committed suicide. It had set itself steadfastly towards comfort and ease, a balanced society with security and permanency as its watchword, it had attained its hopes—to come to this at last. Once, life and property must have reached almost absolute safety. The rich had been assured of his wealth and comfort, the toiler assured of his life and work. No doubt in that perfect world there had been no unemployed problem, no social question left unsolved. And a great quiet had followed.

It is a law of nature we overlook, that intellectual versatility is the compensation for change, danger, and trouble. An animal perfectly in harmony with its environment is a perfect mechanism. Nature never appeals to intelligence until habit and instinct are useless. There is no intelligence where there is no change and no need of change. Only those animals partake of intelligence that have to meet a huge variety of needs and dangers.

分析：

这一个例子同样选自《时间机器》，这一段落描述了时间旅行者目睹了公元802701年时的人类社会以后的感想，具有一定的哲理性与抒情性。

译文：

我一想到人类的智慧梦想是多么短促，就十分悲伤。这个梦想已经自杀了，因为它不懈地追求舒适和安逸，追求一个把安全与永恒当作标语的平衡社会，它实现了自己的希望——终于实现了这个希望。曾经一度，生命和财产几乎已经达到绝对的安全。富人的财富和舒适得到了保障，劳苦者的生活和工作也得到了保障。毫无疑问，在那个完美的世界里，没有失业问题，没有待解决的社会问题。于是世界就变得太平无事。

但我们忽视了一条自然法则，即人们各方面的才智是对变化、危险和麻烦所做的补偿。一只同环境完美协调的动物就是一台完美的机器。只在习惯和本能无用的时候，自然才会求助于智慧。没有变化和不需变化的地方，就不会有智慧。只有那些要遭遇千难万险的动物才能拥有智慧。

第4节　科技论文的翻译

4.1　科技论文的文体特点

科技论文（scientific paper）是正式程度最高的科技文体，其主要目的是用于同行之间的专业交流，也就是论文写作者与阅读者均是专业人士，属于专家写给专家看的（expert-to-expert writing），因而具有科技文体的典型风格，即：语义客观、用词正式、语气正式、结构严密、逻辑性强以及程式化。具体体现在，大量使用专业术语，而不会用普通词汇去指代，例如：在生物或医学领域里，作者会用“leukocyte”指“白血球、白细胞”，而不会用“white blood cells”，用“ovum”指“卵子、卵细胞”，而不用“egg”，因为后者是针对非专业人士用的普通词汇。此外，名词化结构和大量使用长句及包含各种从句的复合句使句子的信息密度极大，同时，多使用第三人称主语的句子或者被动语态，以使文章强调客观推理，不会造成主观臆断的印象。而且，科技论文还具有行文程式化的特点。程式化是指同类语篇大致相同的体例和表达方式。例如，期刊论文的体例依次为：①Title（标题）；②Abstract（摘要）；③Introduction（引言）；④Materials and methods（材料与方法），或 Equipment and test/Experiment procedure（设备与试验/实验过程）；⑤Result（结果）；⑥Discussion（讨论）；⑦Summary结论（概要）；⑧Acknowledgments（致谢）；⑨References（参考文献）。

4.2　科技论文翻译标准

科技论文是非常正式的书面文体，一般而言，是属于专家写给专家看的（expert-to expert writing），要求概念准确、严谨周密、句式严整、行文简练、重点突出、逻辑性强。因此，为了达到准确传递信息、维护经验交流的目的，对科技论文的翻译，要求做到准确规范、连贯顺畅、简洁明晰、逻辑严谨。具体说来，即选词精准，不能有误解；句型结构严谨，不能产生歧义；逻辑缜密，不能留有漏洞。译文多用正式词、庄重词，不用俚语、俗语；多用具体的词，不用抽象模糊的词；多用客观朴实的词，不用带主观感情色彩的词，更少用比喻、夸张等修辞手段。表达方式上，科技文体力求文字简练顺畅，一目了然。

（1）准确规范。

所谓准确，就是完整地传达原文的全部信息内容，尤其是词义的准确辨析和选定，句子结构的准确理解和处理。所谓规范，就是译文要符合所涉及的科学技术或某个专业领域的专业语言表达规范。要做到这一点，译者必须充分地把握原文所表达的内容，辨析原文词汇，分析句子结构，理解科学内容。科技翻译中的错误或是偏差，会给科学研究以及生产等造成巨大的影响。因此，译者必须根据各专业情况，精确传达原文的信息。例如：

The difference in energy level between the substrates and products is termed the change in Gibbs free energy(AG).

译文 1：

在底物和产物之间能量水平的差异称为 Gibbs 自由能（AG) 的变化。

译文 2：

底物与产物之间的能级差叫作吉布斯自由能（AG) 的变化。

分析：

译文 1 中译者将 energy level 译成“能量水平”显然是错误的，从化工词典和有关教科书中我们发现应译作“能级”。缺乏相关的专业知识，是造成误译的最主要原因。

（2）连贯顺畅。

所谓连贯顺畅，是指译文的语言符合译入语的语法结构及表达习惯，容易为读者所理解和接受。也就是说，译文语言应该明白晓畅、文理通顺、结构合理，没有死译、硬译、语言晦涩难懂的现象。例如：

Modem scientific and technical books, especially textbooks, require revision at short intervals if their authors wish to keep pace with new ideas, observations and discoveries.

译文 1：

现代科技书籍，特别是教科书要求作者不定时地修改内容，如果他们希望与新概念、观察到的事实和发现同步发展的话。

译文 2：

对于现代科技书籍，特别是教科书来说，要是作者希望自己书中的内容能与新概念、新观察到的事实和新发现同步发展的话，那么就需要每隔较短的时间，将书中的内容重新修改。

分析：

译文 1 照搬原文，逐字死译，不符合汉语的表达方式和习惯。而译文 2 中译者通过顺逆结合，主次分明地对全句进行综合处理，采用恰当的词语增译，如“新概念、新观察到的事实和新发现”使译文更通顺。

（3）简洁明晰。

简练是科技文体的又一特点，目的是直陈事理，突出重点，避免抽象笼统的表达形式。

这就要求在翻译过程中，译文要简短精炼、一目了然，尽量避免烦琐、冗赘和不必要的重复。例如：

Irrespective of the approach, it is important to agree tariffs before investments proceed.

译文 1：

不论采用哪种方法，在投资前达成一致同意的价格是重要的。

译文 2：

不论方法如何，重要的是投资前达成一致同意的价格。

分析：

译文 1 也将原文的基本意思表达出来了，但并不符合科技文体言简意赅的特点。译文 2 则更为精练，符合科技文体严肃的文风。

（4）逻辑严谨。

一篇科技文章，词与词之间、句子与句子之间、段落与段落之间除了用表面的关联词等进行衔接外，还有其内在的逻辑关系。此外，汉语和英语在句型结构和表达方式上多有不同，单纯地对句子表层词汇进行翻译和拼凑，既无法使原文与译文形式上对等，也无法体现原文句子的深层关系，如语法关系和主题关系。由于汉英两种语言在逻辑关系和思维推导方式上存在各自的特点，在进行科技翻译时，就必须掌握文章的整体思路和发展脉络，这样才能正确把握原文的意思。只有不受原文句子表层结构的制约，才能翻译出逻辑严谨、准确的译文。例如：

The pressure of increasing population leads to the vertical growth of cities with the result that people are forced to adjust themselves to congestion in order to maintain these relatively artificial land values.

分析：

对原文句子成分间逻辑关系的理解直接影响翻译的准确度。如果认为“in order to maintain...”是修饰靠近它的从句谓语“are forced to adjust...”即“人们要被迫适应拥挤以维持人为的地价”，显然不符合逻辑。实际上，这个不定式短语应修饰前面主句谓语“leads”的。

译文：

人口增长的压力造成城市向高空发展，从而人为地提高地价，其结果是人们被迫适应城市的拥挤状况，住进了高楼。

（5）语言得体。

原语的一个词、一个词组、一个句子，甚至一个段落，在译入语里可能有几个同义而结构不尽相同的语言形式。这样，在翻译过程中，就有一个选择的层次。选择除了涉及语言结构因素（如词性、词语搭配、上下文）之外，还与文体有关。选择的目的是使译文的语言得体，即得相应的文体之体，从而获得较好的翻译效果。

每种文体都有某些遣词造句上的常规。要想整篇得体，必须使译文的词、词组、句子以至段落得体。从等值概念上讲，只有在词、词组、句子、段落等深层意义一致和文体修辞功能相当的基础上，才能最终达到篇章的等值，所以篇章的得体，植根于各语言平面的得体。

词的得体主要在于同义词的选择。同义词中可以区别出不同的感情色彩和文体风格，如词义的褒贬、词的通俗或正式、词的口语或书面语色彩、日常用语或专业用语，等等。同义词产生的最主要原因，是人类对客观世界的认识在不断深化。为了精确地表达意义的细微差别，便有了许多同义词。译者要精确地表达思想，就必须特别注意同义词的差别。就科技论文而言，一般要选择词义严谨的词、符合专业特征的词、与上下文能匹配的词。例如：

This kind of material has been made of great value.

译文：

这种材料已被弄成具有很大价值的东西了。

分析：

“弄”是个日常用词，词义范围很宽。科技文章要求词义明确，在这里不宜使用。

改译：

这种材料经过加工已具有很大价值。

在英译汉中，英语的一些语法词组可译成汉语的成语、俗语等，这样做是因为英语词组中词与词的修饰关系已彼此渗透和融合，词组的意义并不等于各单词意义的简单相加。如果对等地译成汉语词，并将其置于同样的搭配关系中，那就会造成病句和意义的丢失等后果。

句子是表示一个完整意思的最小语言单位。在言语交际中，句子是十分重要的。句子的功能转换以词和词组的功能转换为基础，又要兼顾段落、篇章的和谐统一。在词、词组、句子、语段、篇章这五个主要语言平面中，句子是承上启下的关键平面，也是翻译过程的主要着力点。翻译者必须使译文句子得体，以求得更大层面上段落、篇章的得体。下面我们就同一英文句子的四种不同的汉语译文，来说明得体与否以及选择的成败。例如：

In practice, the selected interval thickness is usually a compromise between the need for a thin interval to maximize the resolution and a thick interval to minimize the error.

译文1：

层的厚度的选择是相当困难的，因为这实际上存在着难于调和的矛盾：从提高分辨率的角度来考虑，总希望把层选得尽可能薄；但从减小误差的角度来考虑，又总希望把层选得尽可能厚。那么，究竟要选取多大的厚度才算合适呢？这就往往需要在反复权衡其利弊得失之后，才能得出一个最佳的折中方案。

译文2：

实际上，层的厚度的选择，往往需要在下列两者之间得出一个折中方案，即从提高分辨率的要求来考虑，总想把层尽量选得薄一些；但从减少误差的要求来考虑，又总想把层尽量选得厚一些。

译文3：

实际上，所选择的层的厚度通常是可最大限度提高分辨率所需的薄层和可使误差降至最小的厚层之间的平均值。

译文4：

为保证最大分辨率必须选用薄层，为使误差最小却须选用厚层。实际上，通常选择介于两者之间的最佳厚度。

分析：

以上四种译文都能够传达原文的含义，但是就得体而言，还是存在较大差异。

译文1的译者认为，原文“词理本深，难于共喻”。因此通过“前后引衬”，加上“相当困难”“难于调和的矛盾”“反复权衡其利弊得失”等主观色彩较浓的文字，使译文表层结构膨胀，译文句子明显长于原文句子。这种增加信息容量、不顾原文形式的做法，是不得体的。这很容易让读者产生疑惑心理。

译文2虽较译文1有所改进，但与原文的精练相比仍有差距。原文“a compromise”之后有一个很长的介词短语作修饰语，为本句信息中心，但将“compromise”译成“折中方案”放在句子中部，未能达到原文的强调作用。

译文3虽然很短，但过于照搬原文形式，译成了一个很长的句子，读起来并不流畅，也不符合汉语的表达习惯，也并不得体。

译文4畅晓自然，简练通顺，表层结构安排得体，附着信息与中心信息，相得益彰，把中心信息译为“最佳厚度”置于句末，形成末尾焦点（end-focus)，符合汉语表达习惯，是得体的做法。

要达到上述科技翻译的基本标准，译者首先应精通原作的语言，具有较强的外语应用能力。其次，译者还应不断提高自己的中文水平和文化功底，加强专业理论修养和丰富生活经验。此外，严肃认真、一丝不苟的翻译态度，对原译语字斟句酌、精益求精、勤查字典、考究资料的工作方式都是获得好译文的不可或缺的条件。

4.3 翻译实例

科技论文是科技研究人员研究成果的直接记录，包括理论性论文和实验性论文两大类，或阐明科学理论、或描述科学实验。其内容比较专业化，艰深晦涩；文字也均较正规、严谨。科技论文的结构已经形成了较为固定的格式，一般包括标题、摘要、关键词、引言、正文、结论几部分。

（1）标题的翻译。

标题语言精练，主题明确，在结构上以名词短语居多。例如：

New Technology for the Next Generation of Commercial Transports

—Real or Imaginary

分析：

这一标题虽然以名词为中心词，但整个结构却是一个省略的疑问句。

译文：

下一代商用运输机的新技术是现实还是幻想?

Pretension Diagrams for Bolted Joints

分析：

这一标题的中心词为名词，通过介词与后面的定语联系起来。在翻译时必须首先弄清标题的语法结构以及各个词语之间的相互关系，才能准确地译出标题。

也有其他形式的标题，标题的格式并非完全一致，需视表示的含义而定。有时标题过分简练，或含专业术语缩略语，意思一时不易弄清，译者不必拘泥于行文顺序，可读完全文后再译标题，这样更能确保标题翻译的准确性。

译文：

螺栓结合的预张力图

（2）摘要和关键词的翻译。

论文摘要是对一篇论文中心思想以及主要观点的简短而清晰的概括。论文摘要的目的就是让读者概览文章大意，迅速了解文章主旨。一般来说，学术论文的摘要应该具有高度的信息浓缩性、准确性和独立成篇性。一篇中文学术论文的摘要一般最多300汉字，英文论文摘要在150词左右为宜，摘要中不应该包括本学科的常识性内容，当然更不可包括论文没有涉及的内容。

关键词是为了文献标引工作而从论文中选取出来以表示全文主题、主要内容、研究目的等的单词或术语。关键词的选取要能反映论文的主题思想和核心概念。每篇论文可以选用3到8个关键词。

论文摘要和关键词的翻译应做到忠实、准确，特别是专业术语的翻译必须规范，且应和论文作者在论文正文里标注的术语英文名统一起来。在翻译摘要时，如果不能明确摘要中某一句话的确切含义，译者就必须查阅原文来获取准确的信息。例如：

Environmental Impact Assessment of Land Use Changes for Urban Development Planning

Abstract: The environmental impacts of urban plans are difficult to assess quantitatively because of the significant uncertainties. This paper divides the social economic activities into spatial grids, with land use changes then used to represent implementation of the urban development plan. Scenario analyses and the Monte Carlo technology are integrated to simulate the effects of the land use changes. An uncertainty analysis is then used to quantitatively assess the environmental impacts of the urban development plan. The methodology is illustrated by a case study of the surface water environmental impact assessment for the Dalian urban development plan. The study identifies the environmental risk to the main river, the environmental pressure, the extreme environmental impact, the critical pollutants, and the critical sources during the planning year. The methodology systematically identifies environmental impact uncertainties in urban development plans and quantitatively assesses the environmental impacts to give decision support for scientific planning.

Key words: environmental impact assessment; urban plan; land use change; uncertainty analysis

译文：

基于土地利用变化的城市规划环境影响评价

中文摘要：城市规划的环境影响具有显著的不确定性，难以定量评价。该文将社会经济活动落实到空间网格上，以土地利用变化表征城市规划的实施，集成情景分析和Monte Carlo技术来模拟土地利用变化，并结合不确定性分析技术构建了城市规划环境影响的定量评价方法。对大连市发展规划的地表水环境影响评价进行了方法应用，识别了规划年主要河流的环境风险、环境压力、极端环境影响、关键污染物和关键污染源。该方法可以系统识别城市规划环境影响的不确定性并定量评价其影响，为科学规划提供决策支持。

关键词：环境影响评价；城市规划；土地利用变化；不确定性分析

（3）引言的翻译。

引言是一篇科技论文的开场白，它写在正文之前。每篇论文引言，主要用以说明论文主题、总纲。常见的引言包括下述内容：①课题的提出背景、性质范围、研究目的及其重要性。②前人研究经过、成果、问题及其评价，即文献综述。③概述达到理想答案的方法。引言一般不分段落，若论文内容较长、涉及面较广，可按上述三个内容分成三个段落。引言里，作者不应表示自谦，也不能抬高自己、贬低别人。引言中可能会出现大量专有名词，尤其是相关研究者的人名、杂志或机构名称，翻译时要注意译名的准确性与规范性。例如：

Fraud in Medical Research: An International Survey of Biostatisticians

Introduction

The public awareness of scientific fraud has increased remarkably since 1980s when a controversy made front-page news, a controversy about a paper investigated for fraud which had as coauthor a Nobel Laureate. During the 1990s, scientific fraud was disclosed on numerous occasions (Lock, 1996). In fact, it was recently suggested that fraud now is “endemic in many scientific disciplines and in most countries” (Fulford, 1998) and may undermine the credibility of scientific research. However, the clandestine character and consequential lack of reliable information make it difficult to study scientific fraud. The characteristics and frequency of scientific fraud, therefore, are generally unknown, and its impact on medical research is unclear. Biostatisticians routinely work closely with physicians and scientists in many branches of medical research and have unique insight into data. In addition, they have the methodological competence to detect fraud and could be expected to have a special professional interest in the validity of results. Biostatisticians therefore could provide unique and reliable information on the characteristics of fraud in medical research. The objective of this study was to assess the characteristics of fraud in medical research by surveying the members of the International Society for Clinical Biostatistics (ISCB).

分析：

在这个引言中，除了具有科技文体共有的特点之外，在综述过程中出现了一些人名、机构名词这样的专有名词，在翻译时，除了注意译文语言的正式准确之外，还要注意这些专有名词的正确翻译，尤其是机构名称切勿望文生义。

译文：

医学研究中的欺诈现象：对生物统计学家的国际调查

引言

在20世纪80年代，一篇以诺贝尔奖得主为合著者的论文因学术欺诈而受到调查，由此引发一场争议，并成为重要新闻。自此以来，对学术欺诈的公众意识已经显著提高。在90年代，学术欺诈更是受到多次揭露（洛克，1996）。事实上，近来有人甚至暗示欺诈“在大多数国家的许多学科领域里已经蔚然成风”（富尔福德，1998），可能会因此破坏科学研究的公信力。然而，由于该行为的隐秘性以及缺乏可靠信息，人们很难研究学术欺诈行为。因此，学术欺诈的特点及频次仍旧不甚为人所知，对医学研究的影响也不清楚。生物统计学家一般与医学研究的许多分支机构的内科医生及科学家有密切合作，并且他们对数据具有独特的洞察。此外，他们在方法论上有能力检测出欺诈行为，因而他们有望对研究结果的效度具有特别的职业兴趣。本研究的目标是通过在国际临床生物统计学会的会员中开展调查来评估医学研究中的欺诈行为特点。

（4）正文的翻译。

正文是论文的主体，一般包括：①实验部分（实验材料、实验设施、实验程序或实验方法）；②结果；③结果讨论。例如：

Influence of Genetically Modified Soya on the Birth Weight and Survival of Rat Pups

Methods

Animals

One group of female rats of 180-200 g weight (6, in 2 cages) was allocated to the experimental group, and received 5-7 g soya flour/rat/day prepared from **Roundup Ready** soya, added to the **rat feed** for 2 weeks. Another group of females (3) was allocated to the **control group**, but their diet was supplemented with the same amount of soya flour prepared from the traditional soya in which only traces (0.08% +0.04%) of the **GM construct** were present, most likely resulting from cross-contamination. We also introduced a **positive control group** (6, in 2 cages), which had not been exposed to soya flour. Therefore, the females only got the standard laboratory feed without any supplementation.

Experiment

After 2 weeks on the diets all the females were mated with 2 healthy **males** of the same age, which had never been exposed to soya flour supplement. In order to avoid infection of **females**, the sperm count and quality had not been determined. We carried on feeding the respective diets to all females during mating and pregnancy. Upon **delivery**, all females were transferred to individual cages, and the amount of soya flour supplement was increased by an additional gram for every pup born. **Lab feed** and water was available for all animals during the experimental period. When the rat pups opened their eyes and could feed themselves (from 13-14 days of age), the daily dose of soya supplement was increased to 2-3 g for every pup, although all rats had free approach to the soya. All rats ate their soya portions well. After the experiment was finished the organs of some pups were taken out and weighed. The level of mortality was analyzed by the one-way **ANOVA**, using the Newman-Keuls test for share distribution. The pups' weight and its distribution were checked by Mann-Whitney test and Chi-square in **StatSoft Statistica v6.0 Multilingual** (Russia).

分析：

这段论文正文的节选主要介绍了试验方法、实验对象及过程，语言具有客观、准确、清晰、条理性强的特点。文中大量使用被动语态，描述了对老鼠的种种处理，译文也同样要准确表达出每一步骤。除此之外，原文中涉及大量专有名词，如数据分析统计或者检测的方法、公司的名称等，专有名词的翻译务必要准确，要查到行业普遍使用的规范名称，否则目标读者会不知所云；专业术语，尤其是一些普通名词充当的术语务必要结合上下文，正确理解，选择合适的意义传达，例如"delivery"，在此，显然是指怀孕母鼠的"分娩"，而非"交货"，"Lab feed, rat feed"中的 feed 显然是指"饲料"，"Lab feed"译为"实验室饲料"，"rat feed"则译为"鼠粮"。此外，文中提到的数据统计分析软件名称属于专有名词，由于每一个数据统计分析软件都有不同版本，翻译出来读者反而会不知所云，因此，这里采用零翻译的策略，即直接将统计软件的名称抄录下来，读者看了一目了然。

译文：

实验方法

实验动物

一组体重为 180 ~200 克的母鼠（6 只，装在 2 个笼内）被划分为实验组，每一只老鼠每一天吃添加了 5 ~7 克豆粉的鼠粮，连续吃 2 周，豆粉由抗农达公司的大豆制作。另一组母鼠（3）被划分为对照组，她们的饮食里添加等量的豆粉，但这个豆粉是由传统大豆制作的，其中只含有微量的转基因成分（0.08% +0.04%），很有可能是因为交叉污染造成的。我们还引入一个阳性对照组（6 只，装在 2 个笼内），这些老鼠一点豆粉都没有吃过。因此，这些母鼠只吃常规的实验室鼠粮，没有任何添加物。

实验

按以上饮食喂养了 2 周以后，所有的母鼠与 2 只健康的同龄公鼠交配，这两只公鼠从未吃过豆粉辅食。为了避免母鼠感染，对精子的数量与质量没有进行限定。在母鼠的交配期以及孕期，我们继续给所有的母鼠喂养其各自的饲料。母鼠分娩之后，所有的母鼠都转移到单个的笼子里，并且为每一个出生的幼鼠增加 1 克豆粉辅食。在实验期间，所有的老鼠都可以得到实验室饲料以及水。当幼鼠睁开眼睛，可以自己进食时（出生 13 ~14 天后），虽然所有老鼠都可以自由享用豆粉，但每一只幼鼠每一天豆粉辅食量还是增加到 2 ~3 克。所有老鼠都喜欢吃它们各自的豆粉。实验结束后，一些幼鼠的内脏被取出称重。死亡率水平采用单因素方差分析，使用纽曼 - 科伊尔斯检验法了解份额分布。幼鼠的重量及其分布采用曼 - 惠特尼检验法和数据统计分析软件 StatSoft Statistica v6.0 Multilingual (Russia) 中的卡方检验法来测定。

（5）结论的翻译。

结论是科技论文全文的总结，其目的在于总结和突出研究的成果。结论是经过科学实验或论证由实践到认识的完整过程后得出的，必须符合事理和逻辑以及实验情况，更要经得起考验。因此，结论的语言特点是：肯定准确，简明扼要。以下是上述译例的结论部分：

Our data allow us to speculate and presume that the negative effect of GM soya on the newborn pups could be explained by two possible factors. Firstly, it can be the result of the transformation and the insertion of the foreign genes, which could penetrate into the sexual/stem cells, or/and into cells of the fetus, as it was observed by Schubbert et al. (1998).

Secondly, the negative effect could be caused by the accumulation of the GM soya residues. However, no mortality was observed with the female rats, nor with the young pups survived, although they also began to eat the GM soya. It is supposed that the effect could be caused by the first factor.

译文：

根据数据，我们做出如下推测，即转基因大豆对新出生幼鼠的负作用可能由两个因素来解释。首先，这可能是由于基因的变化以及外来基因的插入造成的，正如舒伯特等人（1998）所观察到的那样，外来基因可能会渗入性细胞、干细胞，或者/以及胚胎细胞。其次，副作用也可能是由于转基因豆粉残留物的累积造成的。但是，在母鼠中并没有观察到死亡现象，存活下来的幼鼠中也没有出现死亡的情况，而且，这些老鼠也开始吃转基因大豆。因而，据此推测，这个负作用可能是由第一因素造成的。

第5节　产品说明书的翻译

说明书文体包括各种产品的说明书、操作指南或使用手册、故障排除和维修保养方法等。从工业装置说明书，到一般机械或家用电器的使用说明，以及旅游指南、食品说明、医药说明和服务须知等均属此范围。说明书英文对应的表达有“instruction，manual，specification”等。说明书是生产厂家向消费者介绍说明商品性质、性能、结构、用途、规格、使用方法、保养、注意事项、质量保证、销售范围和免责声明时使用的经济应用文书。

5.1　说明书的文体特点和翻译

产品说明书讲究言简意赅、通俗易懂，应避免繁杂冗长。这些文体特征体现在其用词、句法和语篇上，翻译时也需格外注意这些文体特点的体现。

(1) 词汇特点和翻译策略。

专业技术术语

由于产品说明书涉及各个行业和领域，因此会出现某一技术领域中特有的“行话”“术语”。有一类专业术语词汇含义单一、稳定性强，遇到这样的词汇只要查询专业词典或具有一定相关专业词汇储备，翻译起来并不十分困难；但是，有些词汇看起来是常用词汇，但实际上在此处意思却并非我们所熟悉的，而是被赋予了专业的含义。遇到这种情况，翻译时要特别注意，一定要确切理解其产品技术背景下的含义后才能翻译，切勿简单判断，造成错误。

缩略语

为了使文本简洁，便于记忆，产品说明书中常常出现缩略语词汇。如：temp. (temperature) 温度，DC (direct current) 直流电，AC (alternating current) 交流电，I/O (input/output) 输入/输出端口，HD (Hard Disk) 硬盘，HA (hyaluronic acid) 透明质酸；V (Volt)伏，A (ampere) 安。有时同一个缩略语还可能代表不同的词汇。翻译缩略语词汇

时一定要仔细推敲，查找专业词典，找到符合文义的正确意思。

名词化

产品说明书词汇的名词化是指其大量使用名词、名词短语和名词性前置修饰语结构，使其语言简洁、表意凝练。名词化是把句中的动词或者形容词转换为名词或者名词性词组，从而使名词或名词性词组获得动词或者形容词的意义而具有名词的语法功能。产品说明书内容以客观介绍产品的组成部分、工作原理、操作使用、维护保养等为主，需要在上述各个部分里，提及产品各部件名称，还要在附图上说明各部分名称，因此整个说明过程需要大量使用名词。例如：

During the application of the product...而不是用 When you use this product...

翻译时可将名词化结构转回动词来翻译，比如上句译为“使用该产品时……”

(2) 句法特点和翻译策略。

英文说明书的句式具有较强的科学性和逻辑性，即要求用尽可能简单的句子提供尽可能多的信息，并且所提供的信息具有客观性和准确性。产品说明书常常使用以下句式结构：

被动语态

由于产品说明书的说明对象是各种产品，强调的是事物的发生和存在，而通常并不注重是谁使其发生或存在，所以在英文中常用被动语态或非人称形式作主语等语法手段来体现。但由于汉语不需要用被动句式来体现事物的客观性，所以在翻译时，被动语态常常转译成主动语态。例如：

Attention must be paid to the change of position.

应当注意使用位置的变化。

情态动词

大量使用情态动词如 may, can, should, must, can't, shouldn't, mustn't 是产品说明书的特点之一，这主要是为了突出产品具有的特性、功能，或强调注意事项等。例如：

The device should be used only by physicians trained in diagnostic and the rapeutic catheter procedures.

本装置仅限由接受过导管插入诊断及治疗培训的医生使用。

简单句

产品说明书旨在用简明、准确的语言，客观、如实地介绍和描述商品，因此多用简单句，句式简练，行文通俗易懂。翻译时亦要体现其语言的简洁性。例如：

Many parts of this washing machine are made of flammable plastic.

本洗衣机有许多可燃性塑料零部件。

省略句

产品说明书中省略句十分普遍，这与说明书简洁明了的文体特征息息相关。另外，由于说明书所反映的信息流对象已十分明确：厂家/商家给消费者提供信息，给予解释、指导、警告等，所以用尽量言简意赅的语句表达意思。翻译时可同样使用汉语省略结构进行对译。例如：

Contraindications: None Known.

禁忌症：尚未发现

Shelf life: Twelve month (to be kept under the temperature of −18℃.)

保质期：−18 ℃以下保存12个月。

祈使句

说明书中常常需要表述警告、注意事项、操作要点等要求消费者特别注意的事项，这时通常使用祈使句来表达。说明书中的祈使句式常使用示意清晰明确的动词加宾语的结构，如：Make sure that...（确保……），Check to see that...（检查以确认……），Adjust…（调整……），Remove...（移除……），Connect...（连接……）等。例如：

In order to run the software, put the support software CD-ROM in the CD-ROM drive.

为运行该软件，请将支持软件光盘放到光盘驱动器中。

Never use insect repellents on wounds or irritated skin.

切勿在伤口或发炎的皮肤上使用驱虫剂。

一般现在时

产品说明书用于介绍产品使用方法、操作方式或注意事项，所以时态上多使用一般现在时。例如：

In order to ensure optimum performance, we recommend you to clean the desk light twice a year.

为了保持灯具明亮的照明效果，请半年清洁灯具一次。

(3) 语篇特点和翻译策略。

①分章。

产品说明书的内容包括产品的规格、构造、性能、特点、用途、使用方法以及维修保养等部分。在语篇叙述过程中，会将几个部分作为并列的章节进行布局安排，各章节的标题一般用黑色字体标出，醒目明确。各章节独立存在，基本上没有过渡句和过渡段。但是在独立的章节里，也会借助词汇衔接和逻辑衔接达到篇章内部的连贯，使读者了解句子之间的语义联系，体现科技英语的逻辑推理性和严密性。

②分节。

产品说明书在语篇上的另一个特点是篇章布局频繁分节（一般在各个章节中进行分节）。分节和分段不同，段落是指由一个或多个句子组成的语义连贯、相对完整的语言片段。节是指另起一行，在位置上独立的语言片段。它可以是一个段落，也可以是一个或几个句子，还可以是一个或几个短语，甚至是一个单词。分节的目的是逐条标注出不同的方面、特点、步骤等。有时为了醒目，在每一小节前会用“标记或排版符号”进行分隔。在翻译时，一般我们会还原原文分节，在语篇结构上不做过多的调整。

我们通过下面这则说明书（节选）给大家示例说明说明书中的分章和分节及翻译方法。

Electrical Specifications

Voltage/Frequency:	220 V - 50 Hz
Power/Lamp:	27 W/PL - F 27 W/4P
Lampholder:	GX 10q - 4
Electrical Insulation Classification:	II

Operation Instructions

- This desk light is not waterproof and is only suitable for indoor usage.
- No alterations of any kind should be carried out on this desk light.
- Do not use any voltage exceeding a 10% margin of the specified standard.
- Do not touch the lamp inside or the lamp shade when the power is switched on.
- Do not place flammable material near the desk light.
- In case of operation failure, please switch off the desk light, unplug the power cord and contact your nearest Philips dealer.
- This desk light is suitable for Philips PL-F/4P 27 W compact fluorescent lamps only.

参考译文:

灯具电气规格

电压/频率:	**220 V ~ 50 Hz**
功率/光源:	**27 W/PL ~ F 27 W/4P**
灯座:	**GX 10q - 4**
电气绝缘等级:	**II**

操作规范

- 此台灯为不防水型，只适合在室内使用。
- 不可随意更改产品结构。
- 所有电源勿超过额定电压的10%。
- 台灯通电时，请不要触摸光源和灯罩。
- 请不要将高温易燃物品放在灯具附近，以避免发生火灾、触电。
- 在保修期内灯具电缆如有损坏，请务必交于临近飞利浦指定经销商更换。
- 此产品只适用飞利浦 PL-F/4P 27 W 电子节能灯。

译文评析:

上面例子中的说明文字包含两个章节 Electrical Specifications（灯具电气规格）和 Operation Instructions（操作规范），均用黑体字标出。在 Electrical Specifications 章节中分为4个小节，说明产品的各个规格参数；在 Operation Instructions 章节中分为7个小节（各小节前用“圆点”符号标记），说明使用该产品时的注意事项。译文保留了原格式不做变化，包括原文中的“标记符号”。

③非文字说明。

为了帮助消费者学习和理解操作程序，产品说明书中除了用字体和特殊排版符号突出重点外，还经常采用多种非文字的表述方法，如图示说明、图表、操作程序图等。这些非文字表述方法简洁明晰，一目了然，比文字表述更能吸引读者的注意力。翻译时，一般保留原文格式，对图表中的文字内容加以翻译。

5.2 说明书翻译的原则

（1）客观真实地传达原文信息。

产品说明书是以说明为主，对事物加以解说、介绍，使消费者得到某种知识的书面材料。因此我们在翻译时，应该确保概念、定义等信息的翻译准确无误，避免歧义，实事求是地反映产品的实际情况。产品说明书的语言一般不用带感情色彩的词，避免带有浮夸成分，不要求妙笔生花，不要求风格美化，只要实实在在地把有关产品的各项内容解释说明清楚即可。我们在翻译说明书时，可更多地使用字面对应性及准确性较强的直译，而少用字面对应性较弱、可读性较强的意译。

下面是一则洗衣机的使用说明书：

Many parts of this machine are made of flammable plastic. Never place hot or burning objects on or near the washing machine.

When disconnecting the power cord from the power outlet, always take hold of the plug, and not the wire, and pull free. Never connect or disconnect the power plug with wet hands since you may receive an electric shock.

For really dirty clothing use hot water 40 ℃.

For removal of blood stains use cold water only.

STAIN REMOVAL AND BLEACHING

- Add 1/2 cap per liter of water.
- Soak laundry well in solution for at least 20—30 minutes and wash.
- Rinse thoroughly.

Power source: 220 V/50 Hz

Power consumption: 400 W

Washing capacity: 3 kg

Spin capacity: 3 kg

Water supply pressure: 0.3 kg/cm^2—10 kg/cm^2

Net weight: 30 kg

Dimension: 500 mm × 500 mm × 850 mm

参考译文：

本洗衣机有许多可燃性塑料零部件，绝不要把热的或正燃烧的物体靠近或放置在本机上。

从电源插座拔出电源插头时，不要拉拽电线，务必拿住插头，然后拔出。绝对不能用湿手插上或拔出电源插头，以防触电。

清洗过脏衣物，请用40℃热水。

清洗血渍，请用冷水。

去污和漂白：

- 每千克水用1/2瓶盖去污剂。
- 彻底浸泡衣物20～30分钟，然后再洗涤。
- 彻底漂洗干净。

电源：220 V/50 Hz
耗电量：400 W
洗涤容量：3 kg
脱水容量：3 kg
使用水压：0.3 kg/cm^2 ~ 10 kg/cm^2
净重：30 kg
外形尺寸：500 mm × 500 mm × 850 mm

译文评析：

以上洗衣机说明书客观描述了洗衣机的操作规范要求、使用方法、性能、规格等有关情况。译文保留了原文的语言特色和语篇结构及形式特征，采用客观的词语和表达方式予以呈现。

（2）译文语言言简意赅、通顺紧凑、通俗易懂。

由于产品说明书的作用是向用户介绍产品的信息和使用方法等，消费者阅读说明书的目的是获取和了解产品相关的信息，以此来指导产品消费。可以说，实用性是产品说明书的最大特点。同时，由于消费者并非专业人士，往往对产品所属领域比较陌生。因此产品说明书应尽量避免繁杂冗长，而应采用平实、简洁的措辞。在翻译说明书时，我们要注意保留英文说明书的语言特色，除了要忠实地传达原文信息外，还要力求语言的简洁达意，使说明书的译文能为一般读者读懂和理解。但是要注意，英译汉时简洁并不意味着大白话，还应该注意措辞紧凑，符合说明文字的特点和要求。

请看下面这个例子：

Should you encounter some problems during the installation or use of this computer, please refer to this trouble-shooting guide prior to calling the helpdesk. Look up the problem in the left column and then check the suggestions in the right column.

译文：

安装或使用本机如遇问题，请先阅读疑难问题指南，如不能解决问题，再致电客服部。请在指南左栏中查找问题，在右栏查看所建议的解决方法。

译文评析：

该译文简洁明晰，通顺易懂，断句合理自然。英语的介词是使用最活跃的词类之一，但汉语使用的频率要少得多。翻译时，往往对某些表示方位和时间的介词采取省译。如，译文对原文中的“during, of”等介词进行了省译，使得汉语表达简洁而又紧凑。

（3）专业术语翻译准确到位，确保原文的技术性特点。

我们前面讲过，许多产品说明书介绍的产品专业性较强，常常会出现专业技术词汇。我们在做翻译时，首先要参考相关专业书籍和资料，弄懂说明书中具有技术含量的信息的实质内涵，遇到疑难问题应反复推敲。翻译术语时，一定要找到符合汉语专业术语规范的正确表达，避免造成读者的困惑和误解。对于专业性较强的说明文字，翻译时一定要持有一丝不苟、认真严谨的工作态度，切不可粗心大意，似是而非，漏洞百出。这种对译者严格的要求在翻译药品说明书时体现得尤为明显。例如：

ISOKET

For long-term and/or emergency treatment of angina pectoris.

Composition:

Isosorbide dinitrate 20mg per tablet, with prolonged action.

Indications:

Angina pectoris, especially the so-called "effort angina"; prevention and subsequent treatment of myocardial infarction. Vasomotor angina pectoris, a form induced by central nervous disorder, is not affected.

Contra-indication:

Glaucoma.

Dosage:

One tablet of Isoket retard to be taken with water and without chewing in the morning after breakfast, and one in the evening before retiring; this will protect the patient from attacks of angina pectoris during both the day and the night. If an attack does occur, half a tablet should be chewed and then swallowed, whereby an immediate action is produced.

Note:

No side-effects are to be expected, but some patient may develop the well-known "nitrate headache" which in itself is harmless. In such instances, it is recommended that the dosage be reduced. Because this nitrate effect is subject to tachyphylaxis, the Isoket retard medication need not be interrupted. Security for angina pectoris patients for 8—10 hours. In cases of extreme severity, immediate relief can be obtained by chewing Isoket retard.

Packings of 60 tablets.

Made in Germany

参考译文：

异速凯他（山梨糖醇硝酸盐）

供心绞痛长期或急救使用

【成分】每片含二硝酸异山梨醇 20 毫克。长效。

【适应症】专治心绞痛，特别是“一累就犯”这种类型的心绞痛。预防和治疗心肌梗塞。但对由中枢神经紊乱而引起的血管舒缩性心绞痛不起作用。

【禁忌症】青光眼。

【剂量】早饭后用开水服 1 片，不用咀嚼。晚间睡前服 1 片，这样可以避免患者在白天和晚间心绞痛发作。如心绞痛发作，应咀嚼半片，咽下后即可发挥作用。

【注意事项】一般不会发生副作用。有些患者可能出现常见的、并无大害的“硝酸盐性头痛”。如遇此种情况，可酌情减剂量。因为患者对这种硝酸盐作用易产生快速耐受性，所以不必停药。本品可在 8～10 小时内保护心绞痛患者安全。严重患者咀嚼本品可立即缓解。

【包装】60 片包装

德国制造

译文评析：

以上是典型的一篇医药说明书，介绍了一种抗心绞痛的药物——Isoket 的基本信息和服用说明等。医药说明书的文字涉及专业知识，具有很强的技术性。对于说明书中大量的医药学术语，译文都予以了精准的处理，用对应的、准确的汉语医药学术语表达出来。

首先译文对说明书各章节标题（黑体字）进行了准确的翻译。以下对药品说明书中常出现的章节标题的译法进行总结，供大家参考：Composition(成分)，Description(性状)，Action(功能主治)，Indication(适应症)，Contra-indications(禁忌症)，Precaution(注意事项)，Side effects(副作用)，Dosage and Administration(剂量和用法)，Packing(包装)，Expiry date(失效日期)，Manufacturing date(出厂日期）等。其次，对原文中的药品名称的翻译也做了慎重的考虑，如标题药品名“Isoket”，若按音译可译为“异速凯他”或“异舒吉”，为了使中国读者更明白药品的化学性质，这里除用了上面的音译以外，又添加了药品化学名称的译文——“山梨糖醇硝酸盐”。另外，对于疾病名称，译文予以了准确表达，如 angina pectoris(心绞痛)，myocardial infarction(心肌梗塞)，central nervous disorder(中枢神经紊乱)，vasomotor angina pectoris(血管舒缩性心绞痛)，glaucoma(青光眼)，nitrate headache(硝酸盐性头痛）等。

大家在进行医药说明书翻译时，事先查询相关医学资料或者向专业人士请教都是十分必要的，因为，翻译一旦出现差错，会引起严重后果。

对于任何类别的产品说明书翻译，都应坚持内容与形式的统一。在内容上，译文应完整准确地传达原文的含义，并应符合译入语的表达习惯，做到语义明确清晰、杜绝歧义的情况；在形式上，译文也应考虑到原文产品说明书的结构安排、句法特点等，并且尽可能在译入语中再现原文的形式特征。

此外，各类产品说明书的翻译还需考虑到各自的特点和风格。电器说明书常用图示来演示操作过程，说明电器的零部件；药品说明书中常用分子式的表达形式，以体现药品实验结果的明确性和客观性。翻译电器和药品说明书必须无条件地坚持准确、客观、专业化的原则，不能随便使用模糊化的语言。另外，有些产品说明书除了用于介绍产品信息外，还具有一定的广告的功能，因此在翻译语言上要特别对这方面加以表现。比如，食品说明书翻译会涉及语言的美感功能，文字准确精练的同时还要有一定的艺术性，从而激发起消费者的购买欲望。化妆品说明书翻译应注意语言的可读性和感染力，必要时某些语句会使用较为文学化的语言，以增加对产品功能和效果的渲染。

5.3 翻译实例

下面的实例是一则数码相机的使用说明书的部分节选。

Read this first

Trial recording

Before you record one-time events, you may want to make a trial recording to make sure that the camera is working correctly.

No compensation for contents of the recording

Contents of the recording cannot be compensated for if recording or playback is not possible due to a malfunction of your camera or recording media, etc.

Back up recommendation

To avoid the potential risk of data loss, always copy (back up) data to a disk.

Precaution on copyright

Television programs, films, video tapes, and other materials may be copyrighted. Unauthorized recording of such materials may be contrary to the provision of the copyright laws.

Do not shake or strike the camera

In addition to malfunctions and inability to record images, this may render the recording medium unusable or image data breakdown, damage or loss may occur.

Do not get the camera wet

When taking pictures outdoors in the rain or under similar conditions, be careful not to get the camera wet. Water entering the inside of the camera may cause malfunctions which in some cases may not be reparable. If moisture condensation occurs, see page 143 and follow the instructions on how to remove it before using the camera.

…

Getting started

Charging the battery pack

1. **Open the battery/ "Memory Stick" /CF card cover.**

Slide the cover in the direction of the arrow

• **Be sure to turn off your camera when charging the battery pack (page 17).**

• Your camera operates with the "InfoLITHUM" NP-FRI battery pack (R type) (supplied). You can use only R type battery pack.

2. **Install the battery pack, then close the battery/ "Memory Stick" /CF card cover.**

Make sure the battery pack is firmly inserted all the way, then close the cover.

3. **Open the DC IN jack cover and connect the AC Adaptor (supplied) to the DC IN jack of your camera.**

Connect the plug with the ▲ mark facing left.

• Do not short the DC plug of the AC Adaptor with a metallic object, as this may cause malfunction.

• Clean the DC plug of the AC Adaptor with a dry cotton bud. Do not use the dirty plug. Use of the dirty plug may not properly charge the battery pack.

4. **Connect the power cord to the AC Adaptor and to a wall outlet.**

The CHG lamp lights up when charging begins.

The CHG lamp goes off when charging is completed.

• After charging the battery pack, disconnect the AC Adaptor both from the DC IN jack of the camera and from a wall outlet.

Number of images and battery life that can be recorded/viewed

The table show the approximate number of images and the battery life that can be recorded/viewed when you shoot images in normal mode with a fully charged battery pack (supplied) at a temperature of 25 ℃ (77 ℉).

Recording Medium	NP-FRI (supplied)		
	LCD screen	No. of images	Battery life (min.)
"Memory Stick"	On	Approx. 300	Approx. 150
	Off	Approx. 400	Approx. 200
CF card	On	Approx. 300	Approx. 150
	Off	Approx. 400	Approx. 200

译文：

请先阅读本文

试拍

在拍接难得一次的场面前，您可能需要做一次试拍，以确认相机是否工作正常。

对拍摄的内容不予赔偿

因相机或记录媒体等的故障而无法拍摄或播放时，本公司对拍摄的内容不予赔偿。

建议做数据备份

为避免丢失数据的潜在危险，应将数据复制（备份）在磁盘上。

有关版权的注意事项

电视节目、影片、录影带及其他资料可能具有版权。未经许可录制这些数据将会触犯版权法。

请勿摇晃或撞击本相机

可能会造成故障和无法拍摄影像。此外，有可能使记录媒体无法使用或造成影像数据的丢失和损坏。

请勿使相机受潮

当在室外雨中或类似条件下拍摄时，请小心不要弄湿相机。水进入相机内部可能会造成故障，这种故障在某种情况下可能无法修复。如果发生湿气凝聚，在使用前请参阅第 133 页，按照说明除去湿气。

……

准备开始

给电池组充电

1. 打开电池/“记忆棒”/CF 卡盖

按照箭头的方向滑动打开盖子。

- 给电池组充电时，请务必关闭相机的电源（第 14 页）。
- 您的相机要以“InfoLITHUM” NP-FR1 电池组（R 型）（附件）操作。您只能使用 R 型电池组。

2. 安装电池组，然后关闭电池/“记忆棒”/CF 卡盖。

请确保电池组完全插入，然后关上盖子。

3. 打开 DC IN 插孔盖，将交流电源适配器（附件）连接到相机的 DC IN 插孔.

使▲标志朝左，将插头连接至插孔。

• 请勿用金属物件使交流电源适配器的 DC 插头短路，否则可能会造成故障。

• 请用干棉花棒清洁交流电源适配器的 DC 插头。请勿使用脏污的插头，否则可能无法正确地为电池组充电。

4. 将电源线连接到交流电源适配器和墙壁插座

充电开始时 CHG 指示灯会亮。

充电完成时 CHG 指示灯会熄灭。

• 电池组充电后，将交流电源适配器从相机的 DC IN 插孔和墙壁插座拔出。

可以记录/观看的影像数目和电池续航时间

表中显示的是在一般模式下，充满电的电池组（附件）在 25 ℃（77 ℉）的温度下能够拍摄/观看的影像的大致数目和电池续航时间。

记录媒体	NP-FRI（附件）		
	LCD 显示屏	影像数	电池续航时间（分钟）
“记忆棒”	开	约 300	约 150
	关	约 400	约 200
CF 卡	开	约 300	约 150
	关	约 400	约 200

译文评析：

（1）首先要注意格式。原文中有非常醒目的各章节标题和小节标题，都用黑体标出，其中小节部分还用数字和标记符号的方式进行了标注。这些是非常典型的产品说明书的特征。译文在篇章结构上没有做调整，保持了原文这种一目了然、简洁明快的特点。

（2）术语翻译的准确度对说明书翻译来说至关重要。特别是对于文中出现的一些半术语，翻译时不能想当然地译成我们常见的意思。例如，译文中将原文中的“battery life”译为拍摄时电池的“续航时间”，而不译为“电池寿命”，或“待机时间”（待机时间是指在不开机或不使用的情况下电池能够维持的时间）；“battery pack”译为“电池组”；“jack”译为“插孔”；“cover”译为“盖子”等。

另外，本文在翻译时，对于原文某些专业术语还采取了保留英语原文的处理方法，如 CF（compact flash）卡，不译成“闪存卡”；DC IN 插孔不译成“直流输入孔”；CHG（charge）lamp 不译成“充电指示灯”；LCD（liquid crystal display）screen 不译成“液晶显示屏”等。这样处理是因为这些英文缩略术语都在相机本身及其附件上有标注，用户阅读说明书时对照英文文字可以非常方便地找到相应的位置，以便操作。

（3）原文中大量使用祈使句，这种句式翻译起来难度不大。译文中直接用汉语的无主语的祈使句式进行转化。对于被动句，译文中大多将它们转换为汉语的主动句。如：Make sure the battery pack is firmly inserted all the way. 译为“请确保电池组完全插入。”；The CHG lamp goes off when charging is completed. 译为“充电完成时 CHG 指示灯会熄灭”。这样的翻译更加简洁，符合汉语习惯。

(4) 原文用表格、图示等来显示数据、说明过程，清晰直接，简单明了，这是产品说明书的突出特征之一。译文保留了表格的形式，且表格中的文字尽量简洁，如 Approx. 译为一个单字“约”，而不是“大约”。

(5) 要特别注意，因为原文的页码和译文的页码会有差别，所以原文中标出的页码在译文中有所改动。这虽然看似小问题，但也是翻译工作严谨性的高度体现。

第6节　纪录片字幕的翻译

我们知道，英语国家拍摄的有关科技、历史、自然等方面的纪录片中的解说词或字幕属于科技英语的一种类型，本小节将就英语纪录片字幕翻译的策略和方法予以讨论。

纪录片的字幕翻译是指同步于纪录片画面，出现在屏幕底部的文本，这种翻译可能是从语音（画外音解说词）到翻译文字的转换，也可能是直接对画面英文字幕的翻译。由于源语文本的特殊性，纪录片字幕的翻译具有自身独特的特点。

6.1　纪录片字幕的语言特点

不同题材的纪录片有着不同的语言风格，但它们有一些共同之处，即语言平实简洁，逻辑性强，层次清晰。由于要配合画面的切换，纪录片的语言可能不是连贯完整的表达，因此可能具有明显的思维的跳跃性。从句子结构来看，一方面，为了尽可能严谨地表述，常会使用多重复合句，句子结构会因此而变得复杂；另一方面，由于纪录片解说词本身的特殊性，句子不一定是完整句，可能是省略句，也可能以词或词组的方式出现。另外，由于观看的原因，即使是一个完整的句子，也有可能被拆分成几个片段，在字幕中一条一条展现。由于句中的标点符号按照正常句子结构标注，当句子被拆分成片段，就会出现一条字幕没有标点符号的情况。从词汇方面看，除了人物对话以外，纪录片用词不会使用很口语化的表达，并且经常会出现各种专业术语。

6.2　纪录片字幕翻译的特点和要求

(1) 形式限制。

由于屏幕尺寸以及观看视频时字幕阅读速度的限制，翻译出来的一条字幕在一行内的字数不能太多，一般不超过25个汉字。字幕字体的大小也应该符合多数观众的观看习惯。由于汉字的结构与英文不同，所以中文字幕通常比英文字幕长度要短一些。翻译时对于句子的断句，中文和英文字幕的情况尽量保持一致（即同步性），但标点符号（特别是“逗号”）要遵循汉语标点使用习惯重新标记。

(2) 文本制约。

纪录片字幕的作用是帮助观众更好地理解纪录片要传达的内容，在表达形式上应倾向于使用简单易懂的句子，可以使观众在最短时间内读取内容信息。字幕时间和空间的局限性要求字幕出现与消失都要与画面声音相一致，这就迫使译者在翻译文本处理上更加精益求精，以最短小精悍的句子向观众完整传达纪录片希望表达的内容。

（3）文化差异。

不同语言背后都有着不同的文化背景和语言习惯，因文化差异而产生的表述差异在纪录片字幕翻译中经常出现。因此，在字幕翻译时，一定要对原文中的信息所映射出的与中国文化不同的思维方式、表达方式等予以重视，深刻理解东西方两种文化之间的差异，深层次体会原文的文化内涵，捕捉原语所发出的真实信息，然后用贴合中国人逻辑思维方式的汉语语言形式，准确无误地传达原文内容。

（4）通俗易懂。

由于字幕在纪录片画面中的出现具有短时性和瞬间性，观众一般不可能对其进行反复阅读，因此在做字幕翻译时，为了配合观众的认知和理解，要求字幕翻译通俗易懂，清晰简洁，自然流畅。翻译的字幕要尽量保持与源语文本的语言风格一致，不可拗口。过于烦琐、表达过于专业的字幕翻译会直接影响观众的观赏体验，影响观众的思维连贯性，是不可取的。

6.3 翻译实例

以下是英语纪录片 *Journey to the Edge of the Universe*（《旅行到宇宙边缘》）的英文字幕（节选）和中文字幕翻译对照。为方便文字编排，用“//”表示视频画面中出现的每条字幕的间隔。

Journey to the Edge of the Universe

Our world, warm, comfortable, familiar//But when we look up, we wonder: //do we occupy a special place in the cosmos//or are we merely a celestial footnote? //Is the universe welcoming or hostile? We could stand here forever, wondering//or we could leave home on the ultimate adventure//to discover wonders, //confront horrors, //beautiful new worlds, //malevolent dark forces, //the beginning of time, //the moment of creation. Would we have the courage to see it through//or would we run for home? //There's only one way to find out. //Our journey through time and space begins with a single step. //At the edge of space, only 60 miles up//just an hour's drive from home. //Down there, life continues. //The traffic is awful, stocks go on trading//and *Star Trek* is still showing. //When we return home, if we return home, //will it be the same? //Will we be the same? //We have to leave all this behind//to dip out toes into the vast dark ocean. //On to the Moon. //Dozens of astronauts have come this way before us. //Twelve walked on the moon itself, //Just a quarter of a million miles from home. //Three days by spacecraft. //Barren//Desolate//It's like a deserted battlefield//but oddly familiar. //So close, we've barely left home. //Neil Armstrong's first footprints//looks like they were made yesterday. //There's no air to change them. //They could survive for millions of years. //Maybe longer than us. //Our time is limited. //We need to take our own giant leap. //One million miles, 5 million, 20 million miles. //We're far beyond where any human has ever ventured. //Out of the darkness, a friendly face//The goddess of love, Venus. //The morning star. //The evening star. //She can welcome the new day in the east, //say good night in the west. //A sister to our planet. //She's about the same size and gravity as Earth. //We should be safe here. //But the Venus Express space probe is setting off alarms. //It's telling us, these dazzling clouds, they're made of deadly sulfuric acid. // The atmosphere is choking with carbon dioxide. //Never expected this Venus is one angry goddess. //The air is noxious, the pressure unbearable. //And it's hot, approaching 900 degrees. //Stick around and we'd be corroded, suffocated, crushed and baked. //Nothing can survive here. //Not even this Soviet robotic probe. //Its heavy armor's been trashed by the extreme atmosphere. //So lovely from Earth, up close, this goddess is hideous. //She's the sister from hell, //pockmarked by thousands of volcanoes. //All that carbon dioxide is trapping the Sun's heat. //Venus is burning up. //It's global warming gone wild. //Before it took hold, maybe Venus was beautiful, calm//more like her sister planet, Earth. //So this could be Earth's future. //Where are the twinkling stars? //The beautiful spheres gliding through space? //Maybe we shouldn't be out here, maybe we should turn back. //But there's something about the Sun, something hypnotic, like the Medusa. //Too terrible to look at, too powerful to resist. //Luring us onwards on, like a moth to a flame. //Wait, there's something else, obscured by the sun. //It must be Mercury. //Get too close to the sun, this is what happens. //Temperatures swing wildly here. //At night, it's minus 275 degrees. //Come midday, it's 800 plus. //Burnt then frozen. //The MESSENGER space probe is telling us something strange. //For its size, Mercury has a powerful gravitational pull. //It's a huge ball of iron, covered with a thin veneer of rock. // The core of what was once a much larger planet. //So where's the rest of it? //Maybe a stray planet slammed into Mercury//blasting away its outer layers in a deadly game of cosmic pinball. // Whole worlds on the loose careening wildly across the cosmos. //destroying anything in their path. //And we're in the middle of it//vulnerable, exposed, small. //Everything is telling us to turn back. //But who could defy this? //The Sun in all its mesmerizing splendor. //Our light, our lives//everything we do is controlled by the Sun. //Depends on it. //It's the Greek god Helios driving his chariot across the sky, //the Egyptian god Ra reborn every day, //the summer solstice sun rising at Stonehenge. //For millions of years, //this was as close as it got to staring into the face of God. //It's so far away. //It is burned out, we wouldn't know about it for eight minutes. //It's so big, you could fit one million Earths inside it. //But who needs number? we've got the real thing. //We see it every day, a familiar face in our sky. //Now, up close, it's unrecognizable. //A turbulent sea of incandescent gas. //The thermometer pushes 10,000 degrees. //Can't imagine how hot the core is, could be tens of millions of degrees. //Hot enough to transform millions of tons of matter//into energy every second. //More than all the energy ever made by mankind. //Dwarfing the power of all the nuclear weapons on Earth. //Back home, we use this energy for light and heat. //But up close, there's nothing comforting about the Sun. //Its electrical and magnetic forces erupt in giant molten gas loops. //Some are larger than a dozen Earths. //More powerful than 10 million volcanoes. //And when they burst through they expose cooler layers below//making sunspots. //A fraction cooler than their surroundings, sunspots look black. //But they're hotter than anything on Earth. //And massive up to 20 times the size of Earth. //But one day, all this will stop. //The Sun's fuel will be spent. //And when it dies, the Earth will follow. //This god creates life, destroys it//and demands we keep out distance.

（视频内容来自优酷网）

译前分析：

（1）由于英文字幕多用简单短句、省略句甚至词组构成一个语句单元，很多时候采取直译的方式就可以完成，翻译难度不大。翻译时我们要尽量保持与原文语言风格的一致性，力求译文清晰简洁，通俗易懂。

（2）因为一般情况下中文字幕比对应的英文字幕要短一些，所以将一条英文字幕翻译为中文，字幕长度（汉字字数）问题一般不需要过多考虑。但是，如果遇到特殊情况，一行内字数略多，可以将汉语字幕用两行安排。如："But there's something about the Sun, something hypnotic, like the Medusa." 中文译文为："但是太阳有催眠般的魔力，像美杜莎（古希腊神话中三位蛇发女怪之一）。" 由于译文中有插入的用于解释的话语，语句较长，将字幕插入画面中时，可排成两行。

（3）由于英汉语言断句习惯的差别，译文中的标点符号要根据汉语标点使用习惯重新设置。有时英文中用句号，汉语可以用逗号。另外注意顿号和破折号的使用。

（4）和其他类型的科技语篇翻译一样，纪录片字幕翻译也要注意术语翻译准确性的问题。科技类纪录片通常会涉及一些较为专业的科学观点和理念，自然会出现大量科技术语，翻译时务必要准确把握。如：Venus（金星），Mercury（水星），Venus Express space probe（金星快车空间探测器），molten gas loops（炙热气体环状物（日珥）），sunspot（太阳黑子）等。

（5）注意遇到文化差异现象，汉译时需要恰当地进行转化。

例如：计量单位的差异。英语中习惯用 mile（英里）作为距离单位，而汉语习惯用公里；原文中出现的 degree 指的是华氏温标（Fahrenheit），汉语习惯用摄氏度（℃）表达。所以为了便于观众理解，在翻译时我们有必要进行相应的换算，并将换算后结果用括号备注的方式写在译文中。如："Just a quarter of a million miles from home." 译为"月球距离地球只有 25 万英里（约 40 万公里）"；"At night, it's minus 275 degrees. //Come midday, it's 800 plus." 译为"晚上会到零下 275 度（约 -170 ℃），//正午则超过 800 度（约 400 ℃）。" 另外有一点要说明一下，为了给观众更直观的感受，我们做翻译时所做出的换算值一般用取整后的数值表示，所以并不精确，译文在数值前可以加上"约"字。

又如，原文中几次出现西方文化中有关"神"的名称，这些概念对于普通观众来说比较陌生，不好理解。因此，在翻译时我们可以搭建"文化的桥梁"，将这些我们不熟知的名称或概念加以解释说明，必要时可以用括号的形式备注出来。如："But there's something about the Sun, something hypnotic, like the Medusa." 译为"但是太阳有催眠般的魔力，像美杜莎（古希腊神话中三位蛇发女怪之一）。"；"It's the Greek god Helios driving his chariot across the sky. //The Egyptian god Ra reborn every day." 译为"它是驾着战车穿越天际的希腊美男子太阳神，//是每天重生的埃及神'拉'（埃及神话里的太阳神）。"

参考译文：

旅行到宇宙边缘

我们的世界，温暖、舒适、熟悉，//但当我们仰望天空，我们想知道：//我们是居住在宇宙中一个独特的地方？//或仅仅是太空的小小一隅？//宇宙是友善的还是充满敌意？//我们是一直站在这里猜想，//还是离开故园来一次终极探险，//去发现奇观、//直面恐怖、//美丽的新世界、//邪恶的黑暗势力///时间的起点///创世的时刻。//我们是否有坚持到底的勇气？//或是逃回家？//想找到答案只有一个方法。//我们开始一场时空之旅。//离地面60英里（约100公里）高度，就是太空边缘//仅仅一个小时车程。//地面上，生活在继续，//车水马龙，商业繁忙，//《星际迷航》还在播出。//当我们回到家，如果我们还能回家，//一切还会如初吗？//我们还是原来的我们吗？//我们必须抛掉这一切//步入前方广阔的黑暗海洋，//前往月球。//在我们之前，已有数十名宇航员前往月球，//其中12人踏上了月球。//月球距离地球只有25万英里（约40万公里），//坐宇宙飞船需要3天。//贫瘠、//荒芜。//月球看起来就像一个遗弃的战场。//但是惊人的熟悉，//那么近，就像我们几乎没有离开家。//尼尔·阿姆斯特朗的第一个脚印，//看起来就像在昨天留下的。//那里没有能够改变脚印的空气，//这些足迹会存在数百万年，//或许比人类存在的时间都长。//我们的时间是有限的。//我们必须自己踏出自己的一大步，//100万英里（约160万公里）、500万英里（约800万公里）、2 000万英里（约3 200万公里）。/远征前人未曾到达的地方。//一个友善的面孔从黑暗中浮现。//爱之女神——金星//启明星、//长庚星。//它可以在东方迎接新的一天，//还会在西方道晚安。//金星是地球的姊妹星，//大小和引力与地球相仿。//我们在这里应该是安全的。//但是金星快车空间探测器却时刻提醒着我们，//告诉我们那些耀眼的云层是由致命的硫酸构成的。//金星的大气层充斥着二氧化碳。//没想到金星是一个愤怒的女神。//有毒的空气、难以忍受的压力，//还有炙热——温度接近900度（约500 ℃）。//多待一会，我们会被腐蚀、窒息、压碎和烤焦。//任何东西都无法在这里生存。//即使像苏联金星号机器人探测器，//它的厚重装甲已被这极端的大气环境给毁了。//从地球看她是多么可爱，近观这个女神却是可怕的。//她是来自地狱的姊妹。//数千座火山犹如长满了痘疮。//大气层中的二氧化碳留住了太阳热量。//金星正在燃烧。//这是无节制的全球气候变暖。//在它变暖之前或许是美丽而平静的，//很像她的姊妹星——地球。//也许就它是我们地球的未来。//那些闪烁的星星哪里去了？//美丽的圆球滑进了太空？//或许我们不应该来这里，我们应该回去。//但是太阳有催眠般的魔力，像美杜莎（古希腊神话中三位蛇发女怪之一）。//可怕得让人无法正视，也无法抗拒她的力量。//等等，这里还有其他东西被太阳炙烤。//它一定是水星。//太靠近太阳，就会这样//这里温度剧烈变化。//晚上会到零下275度（约－170 ℃），//正午则超过800度（约400 ℃）。//烧焦再被冰冻。//信使号空间探测器发现了一些奇怪的现象。//相对体积来说，小小的水星具有强大的引力。//它就像一个裹着薄薄岩石层的大铁球。//这是一颗大行星留下的核心。//其他部分去哪里了呢？//或许是一个迷路的行星猛地撞进了水星，//在一场致命的宇宙弹球游戏中水星的外层被炸掉了。//这些游荡的行星在宇宙中疯狂地掠过，//毁掉他们道路上的一切。//我们就在其中，//脆弱、裸露而且渺小。//这一切都在告诉我们该回去了。/但是谁又能抗拒得了//散发着迷人光彩的太阳。//我们的光线，我们的生命//我们的一切都被太阳控制着，//依赖于它。//它是驾着战车穿越天际的希腊美男子太阳神，//是每天重生的埃及神“拉”（埃及神话里的太阳神）//以及巨石阵夏至的日出。//数百万年来，//我们对太阳神的景仰止于远观。//由于实在太远，//如果太阳熄灭了，8 分钟后我们才会知道。//太阳大到可以装下 100 万颗地球。//谁需要这些数字呢？我们看到了它的真面目。//我们每天看到太阳挂在天空的熟悉面孔。//现在近观它，又陌生起来。//由炙热气体形成的汹涌大海，//表面温度超过 1 万度（约 5 000 ℃）。//难以置信的是它的核心温度则可以到数千万度。//这里热到足以能够//每秒钟把数百万吨物质转化成能量，//超过人类有史以来产生的所有能量。//地球上所有的核武器对于它都是小巫见大巫。//在地球，这些能量是我们利用的光和热。//近距离观看时，却是令人感到不安。//太阳的电磁活动迸发出巨大的炙热气体环状物（日珥），//有的足有一打地球那么大。//释放的能量超过一千万个火山。//当它喷发的时候，就会露出下面温度较低的部分，//形成太阳黑子。//太阳黑子比周围温度低一些，所以看起来是黑的。//但仍比地球任何东西都热。//太阳黑子同样巨大，超过地球大小的 20 倍//但是总有一天，这一切都会结束。//太阳的燃料会耗尽。//太阳死去，地球也会随之死亡。//这个神祇创造了生命，也会摧毁生命，//要求我们保持距离。

第 7 节　练习

一、思考题

1. 什么是英语科技新闻？它具有什么风格特点？
2. 英语科技新闻的标题具有什么特点？翻译时应该如何处理？
3. 英语科技新闻具有什么写作风格？翻译时如何处理以体现出这些风格来？
4. 科技文体依据其正式程度，有何划分？科普语篇有何特点？
5. 科普语篇的翻译应遵循什么原则？
6. 科幻小说有何文体特点？应遵循什么翻译标准？
7. 科技论文的文体特点是什么？
8. 科技论文的写作程式是什么？
9. 翻译科技论文时应该遵循什么原则？
10. 产品说明书与其他科技文体相比较，有什么特别的文体特征？
11. 产品说明书的翻译的指导原则有哪些？翻译时有什么需要特别注意之处？
12. 说明书文体都包含哪些类别？对于各种不同类别的说明文体，在译文的处理上有没有不同之处？
13. 英语记录片解说词或字幕文本与其他类型的科技语篇相比在语言上有什么特点？
14. 英语记录片字幕文本英译汉过程中需要注意什么？

二、翻译实践

“英语科技新闻”部分

1. 翻译以下英语新闻标题

(1) Edinburgh Zoo's Penguin Cam Proves a Snowy Hit

(2) Panasonic to Release Handheld Game Player

(3) “Hacktivists” Retaliate Against Security Expert

(4) Geely Purchasing Volvo from Ford

(5) China leads Asia in research

(6) 1/3 US Adults to Have Diabetes

(7) Cut carbs, live longer

(8) Dell to shed low-price image

(9) Scientists finish first sea census

(10) GM recalls 341 SRXs

2. 翻译以下英语科技新闻

(1)

Dell to Shed Low-price Image

Dell Inc is getting a marketing makeover, trying to shed its image as a cheaper alternative to rivals, Reuters reported Thursday. The world's No 2 PC maker said it will spend “hundreds and hundreds of millions” on an advertising campaign for its consumer business, to coincide with the launch of new products. “We're going to stop mentioning price as the single most important aspect,” said Paul-Henri Ferrand, the chief marketing officer for Dell's global consumer, and small and medium business division. He said the company wanted to emphasize its premium products, as well as cheaper options.

—Details of the new products, which will include a small laptop that converts into a 10-inch tablet computer, as well as smaller tablet devices, are due to be released Monday.

—The launch comes after Apple CEO Steve Jobs recently dismissed rivals' smaller tablets, saying: “The current crop of 7-inch tablets are going to be DOA, dead on arrival.” Steve Felice, president of the global consumer and small and medium business division, fought back. “Apple is known for dictating what's going to happen in the market place and what a consumer should buy. We're at the opposite end of the spectrum, emphasizing customer choice,” he said. “It's a bit premature to tell customers what they're going to like and what they don't like.”

(2)

Tencent or 360, Not Both

Netizens cannot use the instant messaging software Tencent QQ, unless they uninstall software provided by Internet security company Qihoo 360, Tencent Inc announced Wednesday afternoon on its official website. Those using 360's browser can not access QQ space, a blog on the instant messaging service, either. The move was made to ensure the security of QQ users' account, according to Tencent. A compensation plan will be issued to reimburse paid QQ users who do not want to give up 360 and so cannot use the QQ services they have already paid for.

In response, 360 said it had taken countermeasures to ensure that netizens can use both Tencent QQ and software provided by 360.

(3)

China Leads Asia in Research

Asia, led by China, is fast challenging the US, Europe and Japan in spending on scientific research and development, according to a UN report to be published Wednesday. The UNESCO found Europe, the US and China each contribute 20% of the world's researchers, followed by Japan, with 10%, and Russia, with 7%, AFP reported. From 2002 to 2008, developing countries boosted from 16% to 25% their share of studies published in scientific journals, a benchmark of scientific credibility, it said. China's share alone more than doubled, from 5.2% to 10. 6%. However, Asia still lagged on the 2 key measures of scientific stature and innovation.

(4)

New Way to Regenerate Breasts

US biotech company Cytori Therapeutics has figured out a way to augment breasts using stem cell-enriched fat tissue, local media reported. The discovery means cancer victims with breasts mutilated by surgery, as well as women unhappy with their natural assets, can now grow a new and improved pair, with raw materials harvested from their own body fat. The company believes the same technology can be used to regenerate other damaged tissue, perhaps in the heart, kidneys or bones. It could well be one of the most momentous medical advances of the 21st century.

(5)

DNA That Makes People Stingy

If you have a friend who never buys a round or who rarely pays his or her fair share, try not to get too angry. Because being mean could be in their genes. Scientists have pinpointed a stretch of DNA that makes people stingy with their cash, the Daily Mail of London reported Thursday. Around one of every 4 people carry the "mean gene", which we have inherited from our parents. In the study, those with the gene gave less money to charity than others. Such thrifty people may also constantly borrow cigarettes, rather than buying their own, or regularly borrow cash for bus fares, but rarely pay it back.

"科普语篇"部分

请将下面英文语篇译为汉语。

1.

The champions of hearing, by any standard, are the bats. Bat sounds long went undetected by man because they are pitched two to three octaves above what we can hear. But to a number of bats flying around on a calm, still summer evening—and to the unfortunate moths that can hear them and must try to avoid them—the evening is anything but calm. It is a madhouse of constant screaming. Each bat sends out a series of screams in short pulses, each lasting less than a hundredth of a second.

2.

If ignorance about the nature of pain is widespread, ignorance about the way pain-killing drugs

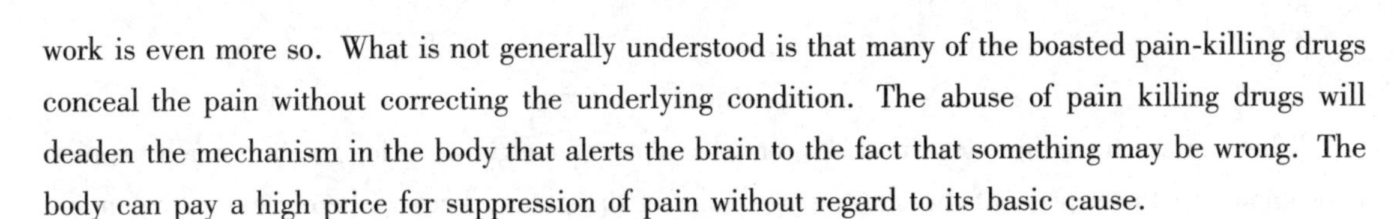

work is even more so. What is not generally understood is that many of the boasted pain-killing drugs conceal the pain without correcting the underlying condition. The abuse of pain killing drugs will deaden the mechanism in the body that alerts the brain to the fact that something may be wrong. The body can pay a high price for suppression of pain without regard to its basic cause.

3.

The Color of the Sky

By Alfred Russel Wallace

If we look at the sky on a perfectly fine summer's day, we shall find that the blue color is the most pure and intense overhead, and when looking high up in a direction opposite the sun. Near the horizon it is always less bright, while in the region immediately around the sun it is more or less yellow. The reason of this is that near the horizon we look through a very great thickness of the lower atmosphere, which is full of the larger dust particles reflecting white light, and this dilutes the pure blue of the higher atmosphere seen beyond.

And in the vicinity of the sun a good deal of the blue light is reflected back into space by the finer dust, thus giving a yellowish tinge to that which reaches us reflected chiefly from the coarse dust of the lower atmosphere. At sunset and sunrise, however, this last effect is greatly intensified, owing to the great thickness of the strata of air through which the light reaches us. The enormous amount of this dust is well shown by the fact that only then can we look full at the sun, even when the whole sky is free from clouds and there is no apparent mist. But the sun's rays then reach us after having passed, first, through an enormous thickness of the higher strata of the air, the minute dust of which reflects most of the blue rays away from us, leaving the complementary yellow light to pass on.

Then, the somewhat coarser dust reflects the green rays, leaving a more orange-colored light to pass on; and finally, some of the yellow is reflected, leaving almost pure red. But owing to the constant presence of air currents, arranging both the dust and vapor in strata of varying extent and density, and of high or low clouds which both absorb and reflect the light in varying degrees, we see produced all those wondrous combinations of tints and those gorgeous ever-changing colors which are a constant source of admiration and delight to all who have the advantage of an uninterrupted view to the west and who are accustomed to watch for those not infrequent exhibitions of nature's kaleidoscopic color painting.

With every change in the altitude of the sun the display changes its character; and most of all when it has sunk below the horizon, and owing to the favorable angels a larger quantity of the colored light is reflected toward us. Especially when there is a certain amount of cloud is this the case. These, so long as the sun is above the horizon, intercept much of the light and color; but when the great luminary has passed away from our direct vision, his light shines more directly on the undersides of all clouds and air strata of different densities; a new and more brilliant light flushes the western sky, and a display of gorgeous ever-changing tints occurs which are at once the delight of the beholder and the despair of the artist. And all this unsurpassable glory we owe to-dust!

4.

Radar

When we are travelling along in a ship or an aeroplane at night or in fog, we are much happier if we know that the captain of the ship, or the pilot of the plane, knows where rocks or mountain tops are, so that he can keep away from them. Until the Second World War, travelers could never have this feeling of safety, because there was no way of "seeing" dangers through fog and cloud. But now there is "radar", a wonderful wartime invention, which has saved the lives of many thousands, both in war and in peace.

Radar is not too difficult a thing for the ordinary man in the street to understand. All of us have heard echoes; we speak in a large hall or before a wall of rock, and our voices come back to us, echoed by the hard substance which they hit, just as light is reflected by a mirror, or a ball is thrown back by the wall it has hit. Scientists know the speed at which sound travels, so they can measure the distance of a wall of rock by making a loud sound and seeing how long it takes to reach the wall and return to the place from which it started. For example, at the same time as the noise is made (perhaps with a car horn) a special watch, which can measure time very accurately, is started. As soon as the echo comes back, the watch is stopped, and it is found that the sound has taken 20 seconds to go and return. We know that the speed of sound is about 1 mile in 5 seconds. So, we know that the noise of our car horn has covered 4 miles, two miles from us to the rock, and two from the rock back to us. The rock, then, must be two miles away.

But sound travels too slowly and cannot go far enough to be useful over long distances or when something is moving very fast. It is therefore no use to an aeroplane, or to a ship which wants to discover a small enemy ten miles away.

Now, it has been known for many years that wireless waves travel at very great speeds: it takes them less than four seconds to go from the earth to the moon and back. But the difficult thing is to measure the time they take to go to a certain distance. If they take less than two seconds to reach the moon, you can imagine that you would not be able to measure (with a watch or a clock) how long they take to go from a ship to the shore, or from an aeroplane to a mountain near it. Radar was made possible by the use of a thing called a "cathode ray tube" (CRT) which can measure millionths of a second. With this tube, we can "see" things at a great distance, and it shows us how far away they are, in which direction they lie, and what movements they are making. On the radar screen we can "see" all around us. The captain of a ship can find his way between rocks and other ships, and the soldiers guarding a town can see enemy aeroplanes hundreds of miles away, and follow them as they approach, perhaps from many different directions at once.

That is radar, one of the wonders of modern science, which is making travel and trade between the nations easier and safer.

"科幻小说" 部分

请将下面科幻小说节选译为汉语。

1.

The thing the Time Traveler held in his hand was a glittering metallic framework, scarcely larger than a small clock, and very delicately made. There was ivory in it, and some transparent crystalline substance. And now I must be explicit, for this that follows—unless his explanation is to be accepted, is an absolutely unaccountable thing. He took one of the small octagonal tables that were scattered about the room, and set it in front of the fire, with two legs on the hearthrug. On this table he placed the mechanism. Then he drew up a chair, and sat down. The only other object on the table was a small shaded lamp, the bright light of which fell upon the model. There were also perhaps a dozen candles about, two in brass candlesticks upon the mantel and several in sconces, so that the room was brilliantly illuminated. I sat in a low arm-chair nearest the fire, and I drew this forward so as to be almost between the Time Traveler and the fireplace. Filby sat behind him, looking over his shoulder. The Medical Man and the Provincial Mayor watched him in profile from the right, the Psychologist from the left. The Very Young Man stood behind the Psychologist. We were all on the alert. It appears incredible to me that any kind of trick, however subtly conceived and however adroitly done, could have been played upon us under these conditions.

2.

About eight or nine in the morning I came to the same seat of yellow metal from which I had viewed the world upon the evening of my arrival. I thought of my hasty conclusions upon that evening and could not refrain from laughing bitterly at my confidence. Here was the same beautiful scene, the same abundant foliage, the same splendid palaces and magnificent ruins, the same silver river running between its fertile banks. The gay robes of the beautiful people moved hither and thither among the trees. Some were bathing in exactly the place where I had saved Weena, and that suddenly gave me a keen stab of pain. And like blots upon the landscape rose the cupolas above the ways to the Under-world. I understood now what all the beauty of the Over-world people covered. Very pleasant was their day, as pleasant as the day of the cattle in the field. Like the cattle, they knew of no enemies and provided against no needs. And their end was the same.

“科技论文”部分

翻译以下科技论文节选

1.

On the Efficient Utilization of Natural Energy Resources in Underground Civil Air Defense Works

Abstract: Based on an analysis of the general features of underground civil air defense works, this paper discusses winter heat storage in underground tunnels. The temporal variations of air dry-bulb temperature, wet-bulb temperature and of wind velocity in an underground civil air defense tunnel in Jinan were examined from both theoretical and practical perspectives. The findings suggest that the use of air in such underground works as a heat source for heat pumps will not only reduce power consumption by the pumps, but also help suppress the attenuation of their heating capacity and thus enhance their coefficient of performance (COP).

Key words: civil air defense projects; energy; air source heat pump; coefficient of performance; energy saving

2.

Model Analysis and Economic Assessment of Solar Shading and Night Time Ventilation Systems in Guangzhou

Abstract: Using DeST, an energy consumption simulation software, we conducted a model analysis of the solar shading and night-time ventilation systems in terms of their application and functioning in the model of a typical commercial building in Guangzhou. 10 proposed solutions were studied in terms of energy conservation and economic efficiency. Results suggest that solar shading and night-time ventilation effectively reduce the load on the building as well as the investment in and cost of air-conditioning. Of the ten proposed solutions, the one which incorporates the controllable inner solar shading and night-time ventilation was found to be the most viable economically.

Key words: solar shading; night-time ventilation; economy

3.

Introduction

The term "genetically modified organisms (GMOs)" refers to plants, microbes1 and animals with genes transferred from other species in order to produce certain novel characteristics (for example resistance to pests, or herbicides2), and produced by recombinant3 DNA technology. There are two standard methods, which are generally used to introduce new DNA (genes) into a plant cell, which is going to be modified. Neither is perfect and both could not guarantee that the rest of the plant genome remains unchanged. Therefore the safety of the GM crops created with the help of these methods cannot be guaranteed for human and animal health, nor for the environment (Ho and Tappeser, 1997; Kuznetsov et al., 2004; Wilson et al., 2004; Ermakova, 2005). Four main of GMOs are now accepted by scientists worldwide: 1) those due to the new genes, and gene products introduced; 2) unintended effects inherent in the technology; 3) interactions between foreign genes4 and host genes5; and 4) those arising from the spread of the introduced genes by ordinary cross-pollination6 as well as by horizontal gene transfer7 (World Scientists Statement, 2000).

4.

Methods

Respondents

The survey was administered in the form of a questionnaire given to 270 students in an elementary physics course at the University of Oslo. This group was chosen for two main reasons: 1) The group was fairly large and easily accessible; and 2) the group mainly consisted of first-year students who had completed secondary education, but had not received formal instruction on radiation since leaving school. This student group can in no way be said to represent the general public. Nonetheless, it can be assumed that the results are indicative of conceptions and attitudes found among those with a general, secondary education in the direction of natural sciences, but no specialisation. Of the 270 students 191 (71%) completed the questionnaire. Of the respondents 53% were women, 45% were men and 2% did not state sex. It cannot be excluded that the 29% who did not respond differed from the respondents in relevant respects. Since this survey does not aim to establish percentages of persons holding this or that conception, but rather aims to identify a few commonly held conceptions of radiation phenomena among non-specialists, the possible bias from non-respondents should not render the results irrelevant.

“产品说明书”部分

请将下面几则英文说明书节选或说明书全文翻译成中文。

1.

Nutrition Information

B-Complex is a combination of B-Vitamins, which helps convert food into energy and maintain normal nervous system function. This product contains 100% of the US Reference Daily Intake (RDI) of folate. Studies show folate helps men and women maintain normal cardiovascular system function. Adequate folate in healthful diets may reduce a woman's risk of having a child with brain or spinal cord birth defects.

Each tablet in this bottle is clear coated for easy swallowing.

2.

Lacovo

Lacovo, scientifically prepared with choice ingredients including malt extract, milk, powder cocoa, fresh butter and eggs, is rich in vitamins A, B, D and organic phosphorous. It promotes health and aids convalescence and is especially good for neurasthenia and mental exhaustion. Take regularly, Lacovo helps build up body resistance against disease. A nourishing beverage for all ages. An excellent gift in all seasons.

For drinking hot: Put two or three teaspoonfuls of Lacovo in a cup, then add hot water and stir until the grains are thoroughly dissolved. Add sugar and milk to taste.

For drinking cold: Put two or three teaspoonfuls of Lacovo in a glass of cold water and stir until the grains are thoroughly dissolved. Then add fresh milk or condensed milk. It makes a delightful and wholesome drink in summer.

3.

Rubber Plug Method of Tubeless Tire Repair

(1) Remove puncture object if still in the tire. Tire is not dismounted from the rim.

(2) Fill tire with air to 30 psi. Dip probe into cement, insert it into injury and work up and down to lubricate injury.

(3) Grasp each end of patch. Stretch and roll center of patch into eye of needle. Remove protective covering from both of the patch, being careful not to touch raw rubber.

(4) Dip perma strip into cement, making sure that all surfaces are coated.

(5) Insert patch slowly and steadily into injury, up to handle. Then turn needle 1/4 turn and remove.

(6) Without stretching the patch, cut it 1/8 inch from the tread.

(7) Inflate to proper pressure. Tire is now ready for service.

4.

Laneige Water Glow Line is created to deliver the beauty of clear water and light. Color your skin with "Face lighting touch" with moist baby lighting instead of high cover makeup.

The effects of brightening and moisturizing make bright and clear skin last longer.

LANEIGE Water Glow Base Corrector SPF41 PA + +

Color correcting—makeup base to have naturally brighten skin tone

Babys skin looks dear and radiant because it has more bluish than adult skin. As we age, skin's bluish color starts to disappear, and the skin tone becomes uneven and looks dull. LANEIGE Water Glows Baby Blue BoostingTM takes out skin's faded bluish color to correct dull, yellowish skin tone and to make skin look more beautiful by expressing clarity like that of baby's skin.

Just as light becomes clearer when more is added, the glow lighting system of LANEIGE Water Glow Base Corrector helps skin become one of the light sources that emits light by itself and makes skin look even and clear.

Experience perfect, even skin tone and light from within skin.

The light, refreshing texture of low viscous essence type applies softly on skin and provides an ideal base skin for makeup. The water-holding layer provides abundant moisture to skin with makeup.

Color Suggestion

20. Rosy Pink: Gives pale skin a healthy complexion.

Micro-prism particles provide the effect of scattering light; apply on areas that need a strobing effect.

40. Light purple: Makes dull, yellowish skin look bright.

60. Light green: corrects reddish skin tone

How-to

Base Corrector is the first step to expressing perfect skin. It provides a bright base by applying on skin before wearing makeup and keeps makeup fresh for long hours. Shake well before use, apply on the face starting from the center and working outward.

Effects & Efficacy: Protects skin from UV rays. (SPF41, PA + +)

Direction & Dosage: Evenly apply an appropriate amount on the face.

5.

CAUTION: Safety Instructions

Use the following safety guidelines to help ensure your own personal safety and to help your computer and working environment from potential damage.

SAFETY: General Safety

Observe the following safe-handling guidelines to ensure personal safety;

- When setting up the computer for work, place it on a level surface.
- Do not attempt to service the computer yourself, except as explained in your Dell™ documentary or in instructions otherwise provided to you by Dell. Always follow installation and service instructions closely.

- To help avoid the potential hazard of electric shock, do not connect or disconnect any cables or perform maintenance or reconfiguration of this product during an electrical storm. Do not use your computer during an electrical storm.
- Do not push any objects into the air vents or openings of your computer. Doing so can cause fire or electric shock by shoring out interior components.
- If your computer includes a modem, the cable used with the modem should be manufactured with a minimum wire size of 26 American wire gauge (AWG) and an FCC-compliant RJ-11 modular plug.
- If your computer has both a modem RJ-11 connector and a network RJ-45 connector, which look alike, make sure that insert the telephone cable into the RJ-11 connector, not the RJ-45 connector.
- Keep your computer away from radiators and heat sources. Also, do not block cooling vents. Avoid placing loose paper underneath your computer; do not place your computer in a closed-in wall unit or on a bed, sofa, or rug.
- Do not use your computer in a wet environment, for example, near a bath tub, sink, or swimming pool or in a wet basement.
- Do not spill food or liquid on your computer.
- Before you clean your computer, disconnect the computer from the electrical outlet. Clean your computer with a soft cloth dampened with water. Do not use liquid or aerosol cleaners, which may contain flammable substances. Allow the computer to dry before reconnecting the power cord to the electrical outlet.
- Ensure that nothing rests on your computer's cables and that the cables are not located where they can be stepped on or tripped over.

CAUTION: Do not operate your computer with any cover (s) (including computer covers, bezels, filler brackets, front-panel inserts, and so on) removed

- PC Cards may become very warm during normal operation. Use care when removing PC Cards after their continuous operation

WARNING: The cord on this product contains lead, a chemical known to the State of California to cause birth defects or other reproductive harm. **Wash hands after handling.**

"纪录片字幕" 部分

请将下面的英文字幕文本翻译成中文。

The animals of Ancient Egyptian King Thuntmose Ⅲ//described a marvelous foreign birds that// "gives birth daily". //Zoroastrians viewed them as spirits//whose cries told of the cosmic struggle//between darkness and light. //Romans brought them on their military campaigns//to foretell the success of future battles. //And today, this bird still occupies an important//though much less honorable position//on our dinner plates. //The modern chicken is descended primarily from the Red Junglefowl, //and partially from three other closely related species, //all native to India and Southeast Asia. //The region's bamboo plants produce//massive amounts of fruit//just once every few decades. //Junglefowl's ability to lay eggs daily//may have evolved to take advantage of these rare feasts, //increasing their population when food was abundant. //This was something humans could exploit//on a consistent basis, //and the birds' weak flight capabilities//

and limited need for space made them//easy to capture and contain. //The earliest domesticated chickens, //dating at least back to 7,000 years ago, //weren't bred for food, //but for something considered less savory today. //The aggressiveness of breeding males, //armed with natural leg spurs, //made cockfighting a popular entertainment. //By the second millennium BCE, //chickens has spread from the Indus Valley to China//and the Middle East to occupy royal menageries//and to be used in religious rituals. //But it was in Egypt//where the next chapter in the bird's history began. //When a hen naturally incubates eggs, //she will stop laying new ones//and sit on a "clutch" of 6 or more eggs for 21 dyas. //By the middle of the 1st millennium BCE, //the Egyptians had learned//to artificially incubate chicken eggs//by placing them in baskets over hot ashes. //That freed up hens to continue laying daily, //and what had been a royal delicacy//or religious offering//became a common meal. //Around the same time as Egyptians were incubating eggs, //Phoenician merchants introduced chickens to Europe, //where they quickly became an essential part//of European livestock. //However, for a long time, //the chicken's reviewed status continued to exist//alongside its culinary one. //The Ancient Greeks used fighting roosters//as inspirational examples for young soldiers. //The Romans consulted chickens as oracles. //And as late as the 7th Century, //the chicken was considered a symbol for Christianity. //Over the next few centuries, //chickens accompanied humans wherever they went, //spreading throughout the world through trade, //conquest//and colonization. //After the Opium Wars, //Chinese breeds were brought to England//and crossed with local chickens. //This gave rise to a phenomenon called "Hen Fever" //or "The Fancy", //with farmers all over Europe//striving to breed new varieties//with particular combinations of traits. //This trend also called the attention//of certain Charles Darwin, who wondered if//a similar selective breeding process occurred in nature. //Darwin would observe hundreds of chickens//while finalizing his historic work//introducing the theory of Evolution. //But the chicken's greatest contribution to science//was yet to come. //In the early 20th century, //a trio of British scientists//conducted extensive crossbreeding go chickens, //building on Gregor Mendel's//studies of genetic inheritance. //With their high genetic diversity, //many distinct traits, //and only 7 months between generations, //chickens were the perfect subject. //This work resulted in the famous Punnett Square, //used to show the genotypes that would result//from breeding a given pairing. //Since then, //numerous breeding initiatives have made chickens//bigger and meatier, //and allow them to lay more eggs than ever. //Meanwhile, //chicken production has shifted to an industrial, factory-like model, //with birds raised in spaces with a footprint//no larger than a sheet of paper. //And while there' s been a shift towards free-range farming//due to animal rights and environmental concerns, //most of the world's more than 22 billion chickens today//are factory farmed. //From gladiators and gifts to the gods, //to travelling companions and research subjects, //chickens has played many roles over the centuries. //And though they may not have come before the proverbial eggs, //chickens' fascinating history tells us a great deal about our own.

（视频内容来自 TED 2020 第 159 期）

参考文献

[1] 蔡基刚. 学术英语（理工）[M]. 2 版. 北京：外语教学与研究出版社，2016.

[2] 陈勇，廖华英. 科技英语阅读教程 [M]. 北京：清华大学出版社，2017.

[3] 陈新. 英汉文体翻译教程 [M]. 北京：北京大学出版社，1999.

[4] 陈桂琴. 科技英语长句翻译方法例析 [J]. 中国科技翻译，2005（3）.

[5] 杜厚文. 汉语科技文体的语言特点 [J]. 语言教学与研究，1981（2）.

[6] 方梦之. 科普与科幻翻译研究的开山之作——郭建中教授《科普与科幻翻译：理论、技巧与实践》序 [J]. 上海翻译，2005（1）.

[7] 郭建中. 科普与科幻翻译：理论、技巧与实践 [M]. 北京：中国对外翻译出版公司，2004.

[8] 高云. 英语科技文章标题的文体与翻译 [J]. 山东外语教学，2004（1）.

[9] 黄海元. 科技英语翻译实用教程 [M]. 北京：国防工业出版社，2009.

[10] 康志洪. 科技翻译 [M]. 北京：外语教学与研究出版社，2012.

[11] 刘宓庆. 文体与翻译 [M]. 北京：中国对外翻译出版公司，2007.

[12] 刘金龙，谷青松. 科技英语阅读与翻译 [M]. 北京：国防工业出版社，2013.

[13] 刘向红，罗晓语. 科技英语文体的名词化结构及其翻译策略 [J]. 湖南工程学院学报，2015（3）.

[14][15] 梁法丽，黄宏. 产品说明书及其翻译 [J]. 天津市经理学院学报，2012（3）.

[16] 李健. 科技英语阅读与翻译 [M]. 北京：外语教学与研究出版社，2009.

[17] 李艳. 英文使用说明书的翻译原则和翻译技巧 [J]. 英语教师，2014（8）.

[18] 李靖，王宇. 基于文化翻译观的英文纪录片字幕翻译 [J]. 林区教学，2016（9）.

[19] 李娜. 纪录片字幕的翻译策略 [J]. 河北广播电视大学学报，2015（5）.

[20] 邱巍，王丽君. 电力科技论文英文标题的翻译 [J]. 沈阳工程学院学报（社科版），2008（4）.

[21] 孙羽，陈莉霞. 词汇衔接手段综述 [J]. 校园英语，2012（8）.

[22] 孙婷. 英文产品说明书的文体特征分析 [J]. 文学教育（中），2010（9）.

[23] 魏海波. 实用英语翻译 [M]. 武汉：武汉理工大学出版社，2009. 7.

[24] 武学民，金敬红. 科技英语翻译中的长难句处理 [J]. 中国科技翻译，1997（4）.

[25] 王宏印. 英汉翻译综合教程 [M]. 西安：陕西师范大学出版社，1989.

[26] 王淑东，张东力，任苗苗. 替代与省略在英汉语篇翻译中的运用 [J]. 吉首大学学报（社科版），2019（3）.

[27] 王振平. 科普著作的文体与翻译 [J]. 上海翻译，2006（2）.

[28] 许建平. 英汉互译—实践与技巧 [M]. 北京：清华大学出版社，2007.
[29] 谢小苑. 科技英语翻译 [M]. 北京：国防工业出版社，2015.
[30] 谢屏. 探索英汉缩略词 [J]. 中国科技翻译，2004 (3).
[31] 相廷礼. 产品说明书的特点及翻译 [J]. 企业导报，2009 (2).
[32] 杨文秀. 科技英语翻译读本 [M]. 南京：南京大学出版社，2012.
[33] 杨丽俊，丁颖. 电视记录片字幕翻译的研究——以《冰冻星球》为例 [J]. 三峡大学学报 (社科版)，2014 (12).
[34] 余高峰. 科技语篇翻译中的连贯策略 [J]. 上海理工大学学报 (社科版)，2012 (3).
[35] 余静. 论纪录片的字幕翻译策略 [J]. 魅力中国，2010 (3).
[36] 张培基. 英汉翻译教程 [M]. 上海：上海外语教育出版社，2009.
[37] 张健. 英语报刊科技新闻文体特点 [J]. 上海理工大学学报 (社科版)，2011 (1) .
[38] 赵萱，郑仰成. 科技英语翻译 [M]. 北京：外语教学与研究出版社，2006.
[39] 周学艺. 英美报刊导读 [M]. 2 版. 北京：北京大学出版社，2010.
[40] 朱庆. 科技英语翻译思维探索 [M]. 北京：国防工业出版社，2007.
[41] 朱永生. 名词化、动词化与语法隐喻 [J]. 外语教学与研究，2006 (2).